Essential Atlas of Ecology

First English-language edition for the United States, Canada, its territories and possessions published in 2005 by Barron's Educational Series, Inc.

Original title of this book in Spanish: *Atlas Básico de Ecología*

Published by Parramón Ediciones, S.A., Barcelona, Spain

Authors: Parramón's Editorial Team

Illustrations: Parramón's Editorial Team

Text: José Tola and Eva Infiesta

Translation from Spanish: Eric A. Bye

All inquiries should be addressed to:
Barron's Educational Series, Inc.
250 Wireless Boulevard
Hauppauge, New York 11788
http://www.barronseduc.com

International Standard Book Number: 0-7641-3094-3

Library of Congress Control Number: 2004107925

Printed in Spain

9 8 7 6 5 4 3 2 1

We are grateful for the generous sharing of graphics by Greenpeace (Nick Hancock/Grace and Beltrá, pages 92 and 93)

WHO (P. Virot, pages 80, 81, 82, and 83) and CESPA (V. San Nicolás, pages 54 and 55)

FOREWORD

This atlas of ecology places into the readers' hands a wonderful opportunity to learn how life forms, plants and animals alike, share and interact with the environment (the soil, air, water, and so forth) and how they evolve and change along with the conditions that surround them. It thus constitutes an extremely useful tool for an in-depth understanding of the physical bases of the ecosystem and the behavior of living creatures in locating and defending a space in which to live and develop, as well as understanding how the great biomes of the planet (the ocean, the forest, the desert, the mountains, and so forth) function. A significant part of this book focuses on the negative influence that some activities exert on the environment and on some ideas to prevent or reduce the damage that is being done to our planet.

The various sections of this book constitute a complete synthesis of ecology. They contain many illustrations and figures that are schematic but accurate, which show the main characteristics of the biosphere, in other words, the space in which life is possible. Such illustrations, which are the nucleus of this work, are complemented by brief explanations and notes that make it easier to understand the main concepts; there is also an alphabetical index that is helpful in locating any subject of interest.

In undertaking the publication of this Atlas of Ecology, our purpose was to create a practical and instructional book that would be useful and easy to use, solidly grounded in science, pleasant, and clear. We hope the readers consider that we have achieved our goals.

TABLE OF CONTENTS

INTRODUCTION

This scene summarizes how an ecosystem works: with light and heat from the sun, grass grows; the gazelle eats it, and the animal in turn becomes food for the eagle; when these creatures die, their remains contribute nutrients to the soil so that the grass can continue growing.

ECOLOGY

This science is dedicated to the study of the relationships that exist between the planet's **life-forms** and the **physical environment** in which they live, plus the relationships that exist among the different types of living beings. Although the earth is very large in comparison with the life-forms that inhabit it, in fact, only a small part of this entire mass is the focus of ecology. This is the **earth's crust**.

Compared with the rest of the earth, the crust is practically a thin layer; sometimes it is said that it is equivalent to the thin skin of an orange. In spite of its small quantity, however, it plays an essential role for all of us. This is where **life** occurs, within a very restricted space. Often the limit is a short distance below the surface of the earth, and most organisms do not move very far away from there. The tallest trees barely exceed 328 feet (100 m), and most of the birds that fly at high altitudes during their migrations are not more than 6,500 to 10,000 feet (2,000–3,000 m) above the ground. However, these birds, just like many other flying animals, spend only short times at those altitudes, because they have to return to the ground to eat and reproduce.

Next we will see some of the main features of ecology, which is a science that perhaps more than any other affects us very directly and generally in all aspects of our daily life. That's the main reason why we pay special attention to what we call **practical ecology**.

Even though the space where life takes place seems very large, it is a thin layer in comparison to the whole earth.

Herbivores, such as the roe deer on the right, feed on plants; carnivores, on the other hand, such as the jackal on the left, feed mainly on herbivores.

BASIC PHYSICAL FEATURES OF THE ECOSYSTEM

Our planet is the environment where we live, and its physical characteristics are the ones that influence our life. We can distinguish three main parts: a solid one, which is the earth's crust; a liquid one, made up of the seas, rivers, and other waters; and a gaseous one, which makes up the air that we breathe. All of this is the physical substrate on which the plants, animals, and other life forms live. The ensemble of this physical substrate and all organisms is called the **ecosystem**.

What we call the ecosystem is always relative, and it depends on what we are studying. In general, we can define our planet as a **global ecosystem**, but within it there are different parts that constitute partial ecosystems. The forests are one, and the seas are another, but we can also speak of the ecosystem of the tropical jungles, the temperate forests, and more. In other words, for the purpose of studying our planet, which is very large for the human frame of reference, it is convenient to divide it up into smaller units.

The ecosystem is not something immovable and permanent, because it is subject to constant changes. These transformations are caused by a continuous supply of **energy** from the sun; this energy is consumed and used in different ways. Another reason is that small portions of **matter** undergo changes, which we refer to as cycles. For example, the minerals in the soil (which constitute fertilizer) pass into the plants, and in the plants these minerals are transformed into leaves, branches, and so forth.

All these materials are the food of **herbivores**, who convert it into meat. Then, when a **carnivore** eats an herbivore, all this matter (which now is called meat but which is made up of plant matter, which, in turn, is made from minerals from the soil) is converted into muscles, bones, and so forth; when the carnivore dies and its body decomposes, all that once again mixes with the soil. In other words, there is a continuous **cycle of matter**.

THE LIVING ECOSYSTEM

We have already seen that the ecosystem is in a state of constant change. That is also true of the organisms that make it up, because their life is limited. We shall see how all these life forms organize at different levels. The **producers** are the beings that are at the closest level to inanimate or inorganic matter (minerals) and use minerals to produce organic matter. These are the **plants**. All other life forms depend on plants, because they are incapable of converting materials into organic matter. These are thus referred to as the **consumers**. They come in two main types: the **primary consumers** eat plants directly (such as cows) and **secondary consumers** (a wolf, for example) eat only meat. In addition to these, there are other consumers that work the process in reverse; that is, instead of constructing more organic matter from food, what they do is decompose the food and turn it into inorganic matter once again. These are referred to as **decomposers**, for example, the bacteria that cause putrefaction.

There is scarcely any life in inhospitable places (very cold, very dry, or with very poor soil); only a few tough plants grow, and a few animals, mainly underground.

LIFE FORMS IN THE ECOSYSTEM

In this section we will see the changes that occur in the relationships among living beings. One of the main types of relationship is what we outlined previously: one uses others as food, so food is one of the principal factors, although not the only one, that motivates living beings. This gives rise to what we refer to as **competition**, which is the struggle between two or more beings over a certain **resource**, whether food, space, a mate, or something else.

In times of drought, Africa's gnus travel hundreds of miles to find fresh pastures.

The living beings that make up part of the ecosystem constitute **populations**, which may involve a single species, such as the zebras of the savanna, or several species, such as herbivores (zebras, buffaloes, gnus, gazelles, and so forth). These populations are not always the same, because they fluctuate in the course of time. In years of great drought, the herbivore population of the African savanna is greatly reduced; but when conditions are more favorable, with plenty of food, that population increases spectacularly.

In addition, the individuals within the populations **move about** fairly regularly. Sometimes this is nothing more than the daily trip in search of food or water, but, in other instances, it involves a long voyage to very distant places—in other words, a **migration**. In the case of the African savanna, this is what happens every year with the gnus.

Selective collection of trash (glass, paper, cardboard, organic remains, plastics, and so forth) makes it possible to recycle many materials.

THE PLANET'S GREAT BIOMES

When we travel to any place on our planet, we encounter different landscapes, some of which are unique, and others are repeated in different places. That general notion of landscape is what is meant by the term **biome**. It is a set of **living beings** organized within the **ecosystem**, with certain important common characteristics. Just as with the ecosystem, which is subdivided into smaller segments, biomes are subject to this type of classification. We speak in very general terms of seas and oceans, of rivers, of forests and jungles, of deserts, of mountains and of polar regions. These are roughly the main biomes that we find on Earth. But within them we can also distinguish smaller units. Under the large heading of the forest biome, for example, we include the forests of the temperate regions, the **tropical jungles** and the **taiga**. Each of these places in turn has its characteristic **fauna**, thereby creating a special ecosystem that can be distinguished from others that appear to be similar.

For ages people have used the forces of nature to produce energy. The photograph shows some old flour mills.

PRACTICAL ECOLOGY

Ecology is a science, but many of its consequences affect us so much in our existence that it becomes a feature of our practical life. What we can call **practical ecology** is the application of the knowledge that comes from **the science of ecology** to further the relationships that we maintain with our planet and the other living beings that inhabit it. This does not involve just the scientists, but all of us. The issues that we deal with in this section are practically all the facets of daily life. The air has become unbreathable in many large cities, rivers look like sewers, and everywhere there are spills of dangerous substances that affect even polar waters or that enter the atmosphere and contribute to the formation of a hole in the **ozone layer**.

In summary, **pollution** is one of the main problems that we encounter today, and it affects not only the health of the planet, but all of us. We will see some of the remedies that exist for these diseases of the natural world. We acknowledge that many of these remedies depend on our behavior—by avoiding the production of unnecessary **wastes**, by **recycling** paper and glass, by conserving water, and by using **alternative energy sources**.

THE BIOSPHERE: WATER, LIGHT, AND ENERGY

The **Earth** is the only planet in the solar system on which it is known that **life-forms** exist. The Earth has a set of characteristics that make the development of **life** possible: appropriate temperature, the presence of water, and so forth. As we study the Earth, we commonly speak of three main layers: the **lithosphere**, which is the solid layer; the **hydrosphere**, which is the layer of water; and the **atmosphere**, the gaseous layer that surrounds the other layers. But we must also remember that living creatures are part of the planet and are extremely important in maintaining its conditions. The set of all living beings forms the **biosphere**, which occupies a narrow band on the Earth's surface.

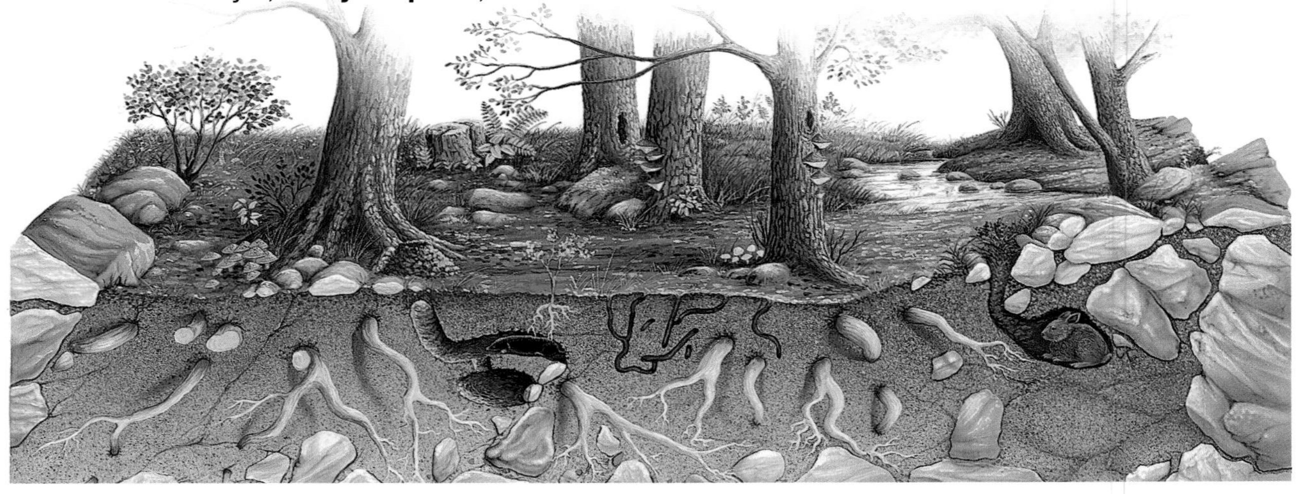

THE BIOSPHERE

In comparison with other layers, the biosphere is very thin, but it exerts a great influence on the other layers (the atmosphere, the hydrosphere, and the lithosphere), because **living beings** interact with the environment that surrounds them and change its characteristics. The thickness of the biosphere is very changeable. In the steppes, for example, it occupies just a few yards, from the deepest areas of the soil reached by the roots of the plants and in which bacteria and fungi develop, to the top of the vegetation, or the point reached by the animals with the largest bodies; in this case, because the vegetation is herbaceous, that is not very high. In the oceans, the biosphere can occupy several miles, from the surface of the water down to the depths where a multitude of deep-sea creatures live.

Seen from outer space, the Earth appears as a blue planet because of the great quantity of water that covers it.

If all the living creatures of the Earth were gathered up and put into a uniform layer on the surface of the planet, that layer would be just $^3/_8$ inch (1 cm) thick.

The biosphere is the space (in the air, the soil, the subsoil, and the water) where conditions are favorable for the development of life.

FACTS ABOUT THE PLANET EARTH

Total surface area	196,860,000 square miles (100%)
Surface area covered by water	139,346,000 square miles (71%)
Exposed land area	57,514,000 square miles (29%)

THE HYDROSPHERE

This is the part of the planet that is occupied by water. The **hydrosphere** includes oceans, seas, rivers, lakes, atmospheric water (clouds and steam), and subterranean waters. **Living beings** depend on water for life. Even the creatures that never drink water in their entire lifetimes need it, and they get it by carrying out certain chemical reactions inside their body. **Life** began in the water, and all living beings have a good part of their body made up of water.

THE DISTRIBUTION OF WATER ON EARTH

Oceans and seas	520,328,000 square miles
Ice (poles and glaciers)	10,036,000 square miles
Freshwater (rivers, lakes, etc.)	88,780 square miles
Subterranean waters	not known

LIVING BEINGS

Living beings are made up of matter, and they need energy to maintain their structure. This energy can be obtained in different forms. **Plants**, for example, use sunlight (**solar energy**). Thanks to this energy they are capable of transforming carbon dioxide in the atmosphere and the minerals from the soil into the organic material that makes up their own body. **Animals** and **fungi**, on the other hand, use the energy produced when the bonds that form the molecules that make up the organic matter used for food are broken (**biochemical energy**).

ENERGY FLOW IN AN ECOSYSTEM

TROPHIC LEVEL

Trophic level indicates every one of the groups of organisms that use energy in the same way: for example, the **producers** (plants) that make organic matter; the **primary consumers** (herbivorous animals), which eat only plants; **secondary consumers** (carnivores), and so forth.

If there were fewer available plants than what the herbivores need (such as the zebras below), they would inevitably die. Similarly, if there were more carnivores (such as the wolf at the right) than herbivores, both would disappear. Thus, ecosystems need balance.

THE ENERGY SOURCE

Solar energy makes it possible for **plants** to grow; plants are the food basis of all **vegetarian animals** (the herbivores) that serve as food for the **carnivores.** As a result, solar light is the energy source for living beings.

ENERGY FLOW

In nature, energy flows in a straight line. Of all the **solar energy** that reaches the surface of our planet, only a small part passes from one **trophic level** to the next one; the rest is lost in the form of **heat** (in the **respiration** of the cells). That is why there are many more **plants** than herbivores (which live off the remaining energy in plants) and many more herbivores than **carnivorous animals**, because they too can make use of only the remaining energy of the herbivores.

In each **trophic level** only about 10% to 20% of the energy stored in the lower level is used.

Only about 0.2% of the solar energy that reaches the surface of the Earth is used by plants.

ATMOSPHERE, SOIL, AND CLIMATE

To understand how **ecosystems** work, we have to take a step back and get to know the inanimate world in which they are located and the processes that take place there. In the case of the planet Earth, the **atmosphere** has a very special composition and structure that made it possible for life to develop. The **soil**, a layer of minerals produced by the erosion of rocks, is the base that supports all land ecosystems. The **climate** is a vitally important factor in determining the type of ecosystem that will develop in any given location.

THE EARTH'S ATMOSPHERE

The **atmosphere** is the gaseous layer that envelops a planet. Its composition depends on several factors, such as the type and the proportion of chemical elements of the planet, the temperature, and so forth. In the case of the Earth, the present atmosphere has a very special composition, which is the product of living beings over millions of years. Molecular **oxygen** (O_2), which currently amounts to 21%, has not always been present. It was caused by the action of the primary filamentous algae, which liberated this gas for millions of years through the process of **photosynthesis**, until today's concentration was reached. This process then reached a point of equilibrium between consumption by respiration and production through photosynthesis.

A PROTECTIVE SHIELD

The atmosphere acts as a barrier to objects that might collide with the Earth's surface. When a piece of rock from outer space penetrates into the atmosphere, friction causes it to disintegrate.

THE COMPOSITION OF THE EARTH'S ATMOSPHERE

Gas	Percentage
Nitrogen	78.00%
Oxygen	20.50%
Argon	0.90%
Carbon dioxide	0.03%
All remaining gases and components	0.57%

Without the atmosphere, **cosmic radiation** would eliminate life and the continual fragments of rock would fill the surface of the Earth with craters.

LAYERS OF THE EARTH'S ATMOSPHERE

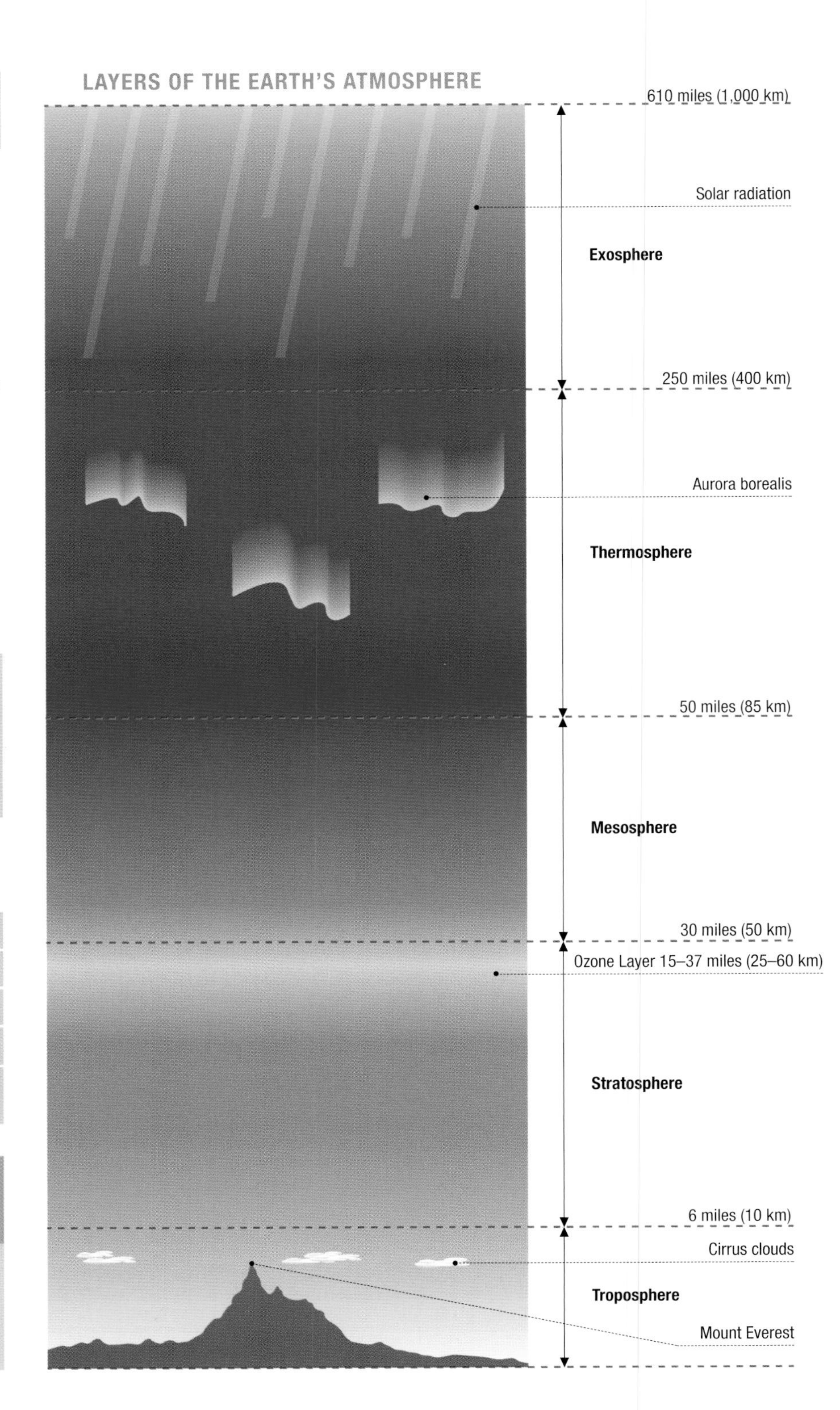

SOIL

The **soil** is the layer of minerals and organic remains created by the erosion of the **rocks** and the action of **living beings**. It is an extremely important element of the **ecosystem**, because the formation of the plant covering that is the base of the **trophic chains** depends on it. In addition, there are nearly invisible **subterranean fauna** that develop in the soil; they are vitally important to the beings that inhabit the surface. Worms, for example, move the earth around and encourage aeration, and thus keep the collected organic matter from rotting and harming the roots of the plants.

When a **forest** is cut or burned, the soil loses its protection against erosion and gets swept away. The disappearance of the soil is very serious, because without it, it is impossible for a new forest to take root.

THE SOIL IS DIVIDED INTO THREE MAIN HORIZONS

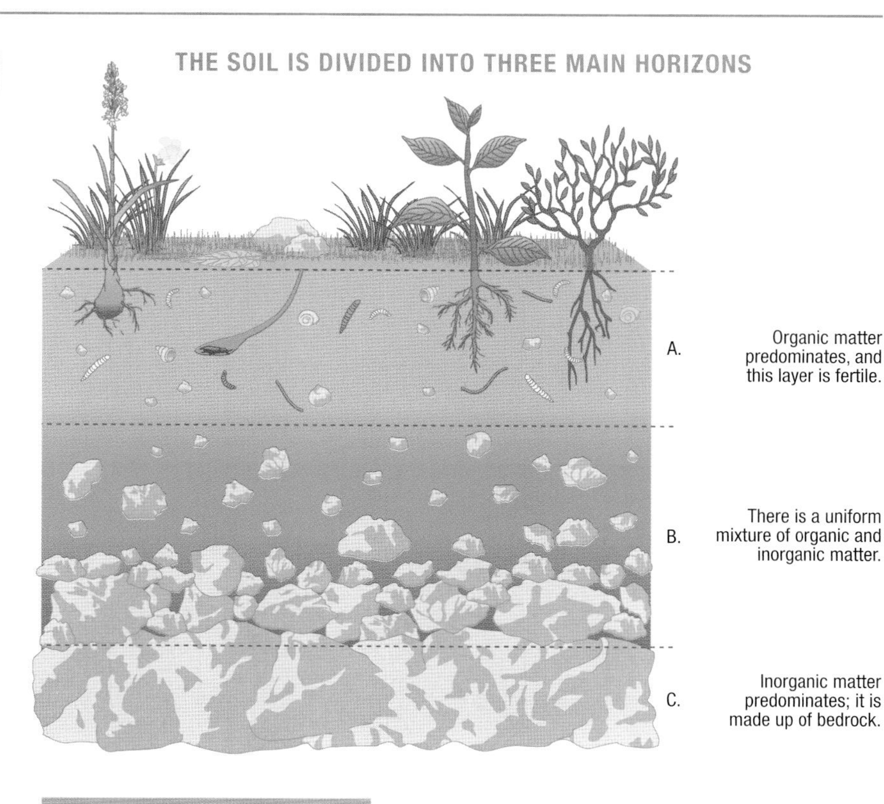

The subterranean fauna oxygenate the soil and facilitate decomposition of the organic matter used by the plant roots.

DECOMPOSERS

These are organisms, such as **mushrooms** and **bacteria**, that break down matter that is difficult for animals to digest and free up essential chemical elements for plant growth.

The jungle dominates hot, rainy areas.

CLIMATE

Climate is defined as the mean value of atmospheric conditions (temperature, precipitation, wind, and so forth) that predominate in a region over a long series of years. It depends mainly on **latitude** and **altitude**. The former is important because the greater the distance from the equator, the smaller the angle of incidence of the Sun's rays, thus, the less heat received. The second factor functions by means of **temperature**, because the greater the altitude, the colder it is. Other important factors that also influence the climate of any given place are the distance from large bodies of water (oceans, seas, large lakes), the presence or the absence of wind, the existence of certain geographic features (mountains, depressions, and so forth), and vegetation.

In northern mountainous regions the climate is cold and the forests tend to be composed of firs.

Climate is very important to living beings, and it forces them to accept its conditions, preventing them from living in certain regions to which they are not adapted.

THE PRINCIPAL CLIMATES

We can distinguish the following climates: **tropical**, **subtropical**, **desert**, **temperate**, **cold**, and **polar**.

THE CYCLES OF MATTER

When a **seed** germinates, a plant begins to grow, and it may turn into an enormous tree. But where does all the matter come from that makes up the trunk, the branches, and the leaves? The seed has a small amount of **reserve matter** that helps the trunk and the root begin to develop, but once this reserve is used up, the plant has to stockpile matter from the surrounding environment and incorporate it into its own structure. In the future, this matter will return to the environment, creating a closed circuit in which the different **elements** spend one period of time in the inorganic environment and another as part of a living being.

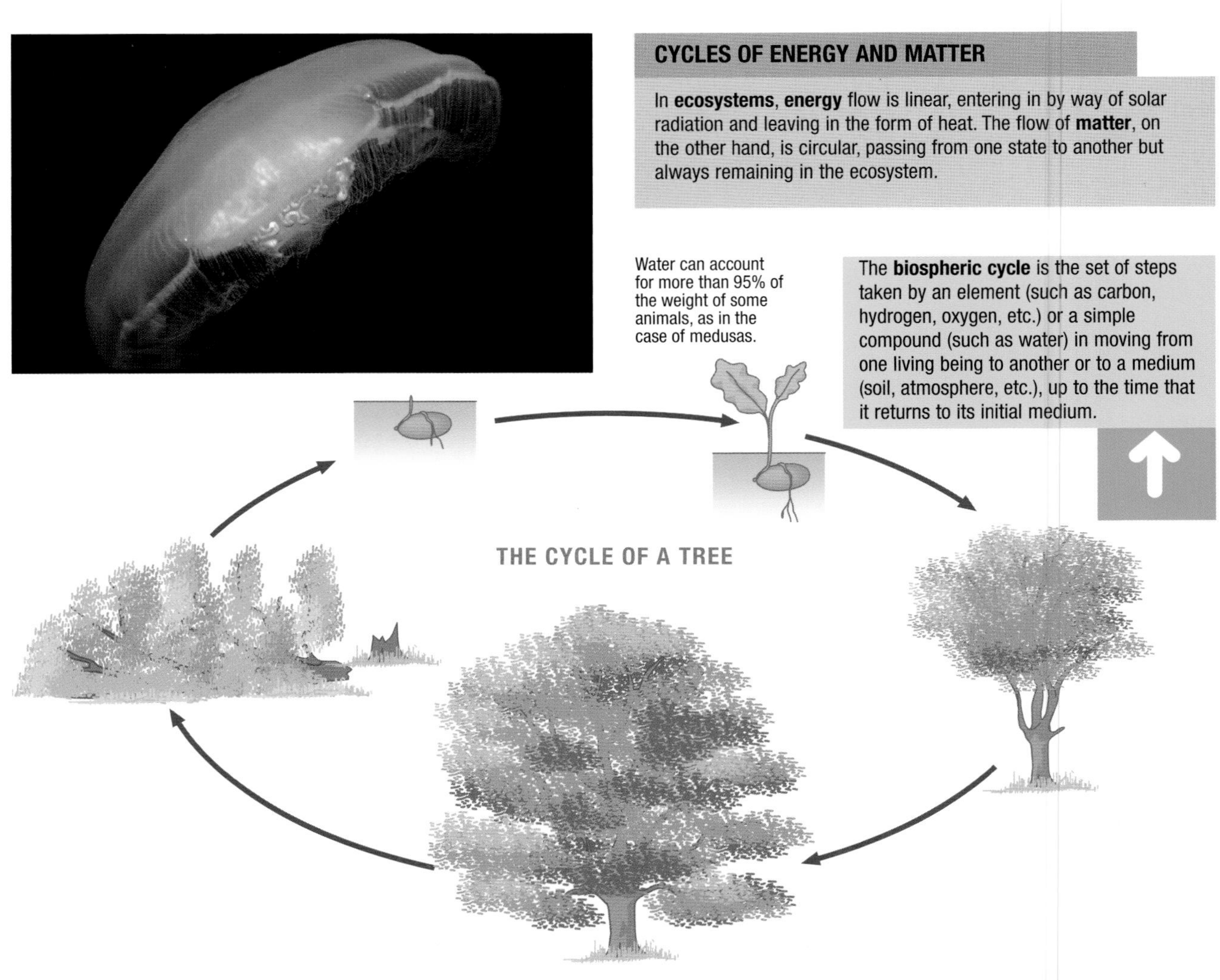

Water can account for more than 95% of the weight of some animals, as in the case of medusas.

CYCLES OF ENERGY AND MATTER

In **ecosystems**, **energy** flow is linear, entering in by way of solar radiation and leaving in the form of heat. The flow of **matter**, on the other hand, is circular, passing from one state to another but always remaining in the ecosystem.

The **biospheric cycle** is the set of steps taken by an element (such as carbon, hydrogen, oxygen, etc.) or a simple compound (such as water) in moving from one living being to another or to a medium (soil, atmosphere, etc.), up to the time that it returns to its initial medium.

THE ELEMENTS THAT MAKE UP LIVING BEINGS

The matter of living beings is made up by quite a number of chemical elements that are obtained from the environment: the **bioelements**. Among bioelements there are a few main ones: **carbon**, **nitrogen**, **phosphorus**, and **sulfur**. Even though water is a molecule and not an element, it is a very important part of the body of animals and plants. All these elements are part of characteristic cycles in which they pass in succession from an inorganic form (soil, atmosphere, river water, etc.) to an organic form (tissue in some living being, part of excrement, and so forth).

THE MAIN ELEMENTS THAT MAKE UP LIVING BEINGS

Carbon C	Magnesium Mg	Molybdenum Mo
Oxygen O	Boron B	Chlorine Cl
Hydrogen H	Iron Fe	Sodium Na
Nitrogen N	Manganese Mn	Selenium Se
Phosphorus P	Copper Cu	Tin Sn
Sulfur S	Zinc	Chromium Cr
Potassium K	Cobalt Co	Vanadium V
Calcium Ca	Silicon Si	Fluorine F

CHEMICAL LIFE CYCLES

It is plants, in fact, that take in most of the **inorganic molecules** from the environment and convert them into the **organic matter** that makes up their own tissues. In turn, **herbivorous animals** feed on them and incorporate those elements into their bodies. The same thing happens with the **carnivores** that feed on the herbivores. In this way, matter can pass from one **trophic level** to another until the elements ultimately return to the inanimate world of nature through the **bacteria** that decompose dead bodies and organic remains. Thus, every element follows a circuit through different parts of the ecosystem.

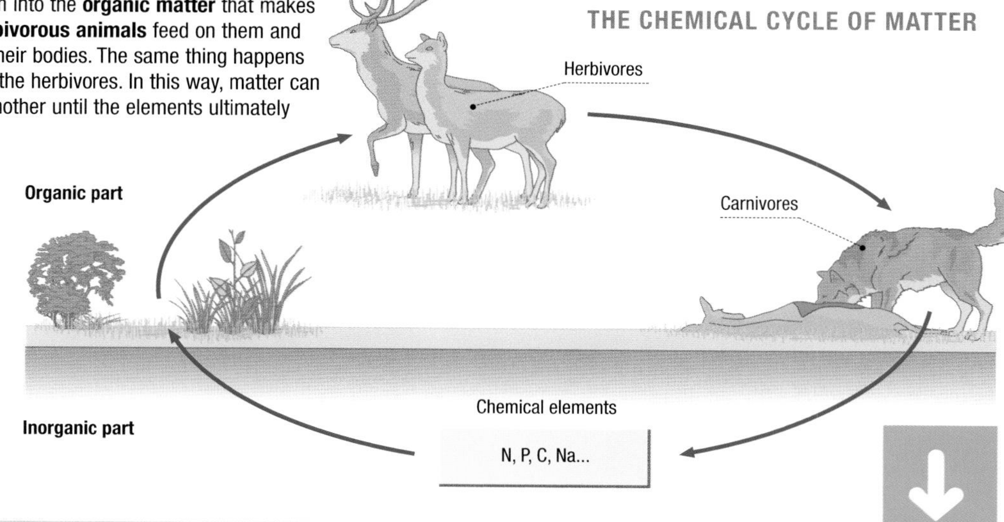

METABOLIC ROUTE

This is the route that a chemical element follows inside the body of an organism, changing from certain compounds into others by means of various reactions.

In some cases, as with nitrogen, phosphorus, and sulfur, the cycles of matter can be very complicated, because many organisms and **metabolic routes** are involved.

THE CARBON CYCLE

CO_2 CO_2 CO_2 CO_2 Respiration CO_2 Respiration CO_2 Meat Glucose Aquatic plants CO_2 Bacteria

CARBON

This is the main element that makes up organic molecules. Approximately half of the solid matter in every living being is made up of this element. Most of the carbon usable by plants is found in the form of carbon dioxide dissolved in the atmosphere (in the amount of 0.03%).

OXYGEN

The natural cycle of oxygen is the opposite of carbon. Thus, photosynthesis releases oxygen into the environment. When animals breathe, they combine oxygen and carbon to produce carbon dioxide, which they expel.

Industrial activity is modifying the balance of the carbon cycle, because it releases great quantities of CO_2 into the atmosphere as a result of petroleum combustion.

THE CARBON CYCLE

Imagine a **carbon** atom that is part of the atmospheric **carbon dioxide** (inorganic form). One day a plant absorbs it through one of its leaves and involves it in the process of **photosynthesis**. It is converted into a molecule of glucose (organic form), which forms a small part of that plant's body. An herbivore eats the plant and thus gains control over the carbon atom. Next, a predator kills the herbivore and eats it, taking the carbon atom into its own body. There it makes up part of the animal's tissues until the time it dies. **Decomposing** organisms transform part of the cadaver into carbon dioxide and release it into the atmosphere. This is the history of one carbon molecule, the **carbon cycle**.

THE DIFFERENT CYCLES

Nitrogen and **phosphorus** are other important elements that make up part of living beings and that are found in nature in various forms. Their cycles commonly are more complex than that of carbon. Even though **water** is not a chemical element, it is essential to life, and its cycle varies greatly. We can encounter it in very different forms in the natural environment and inside organisms, where it may be the most significant component.

THE NITROGEN CYCLE

The Earth's **atmosphere** contains a great quantity of gaseous **nitrogen** in the form of N_2. However, neither plants nor animals can use this form of nitrogen; the only ones that can take advantage of it are a few bacteria and marine algae that transform it into **ammonia** (NH_3). Plants still cannot use this ammonia, and once again it is a group of bacteria that is able to convert it into nitrates (NO_3), which are the only form of nitrogen that plants can use for growth. The plants incorporate this into their tissues, and, in that form, they can pass into the herbivores that feed on them, and later on, into the carnivores, and so forth.

Nitrogen is part of proteins and the nucleic acids (DNA and RNA)

In farm fields, elements are taken out of the system, because people export the plant production to different places. The soil becomes used up because the cycles cannot be completed. That is why it is necessary to fertilize the fields.

The decomposition of animal and plant remains, as well as the urine from animals, releases nitrogen compounds into the environment. Certain bacteria convert them into nitrates so that they can once again be used by plants.

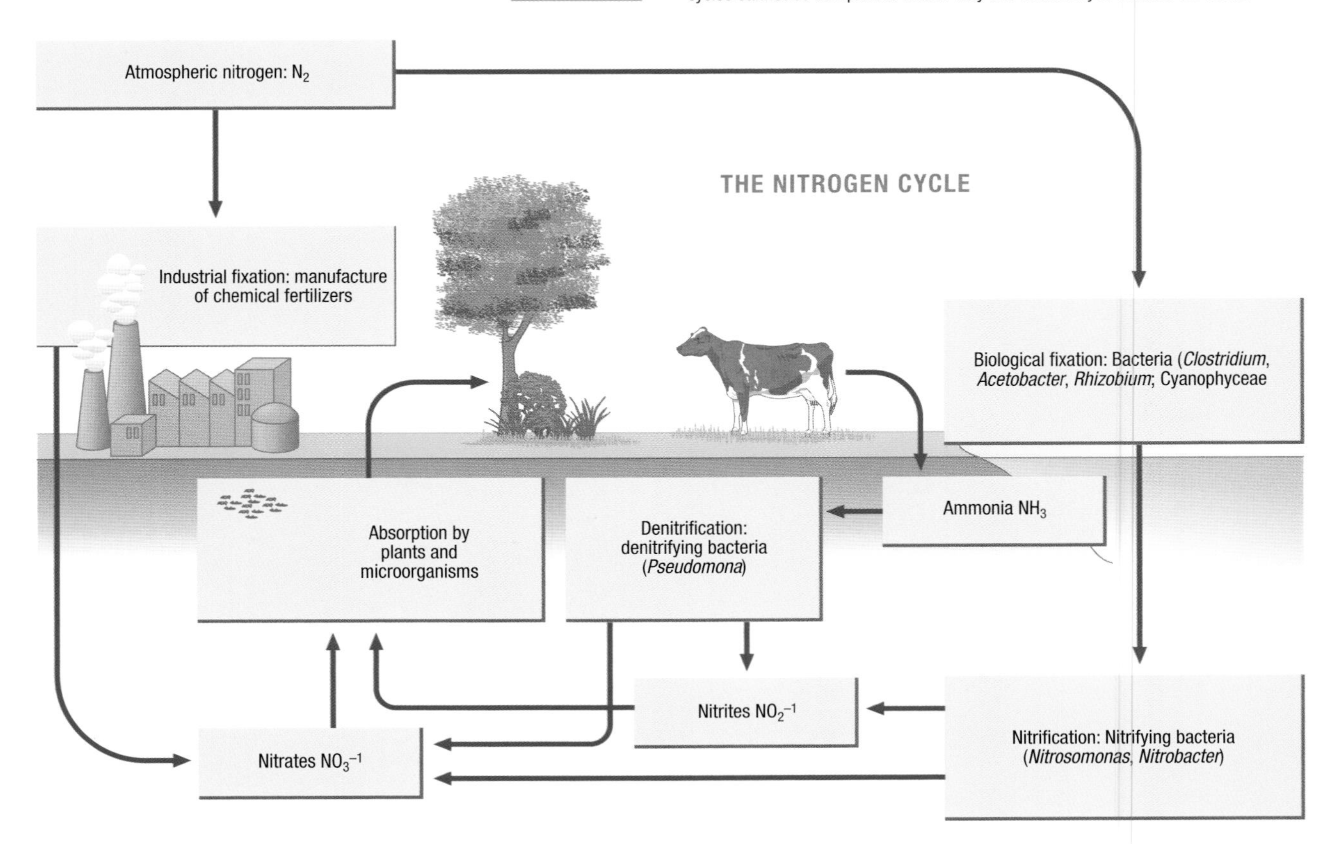

THE PHOSPHORUS CYCLE

Phosphorus is an element that is found accumulated in mineral form in the natural environment. It is freed up when slightly acidic water causes a series of chemical reactions with the rock and forms a compound. This molecule may become part of the **soil**, where it is usable by land plants, or it may reach the sea, where it is used by algae, especially the **phytoplankton**. Once phosphorus is incorporated into the organic matter of living beings, it passes from one level to another through the food chains and is not freed up into the environment until the bodies of those beings decompose.

Marine birds excrete tremendous quantities of phosphorus, and their excrement is a very important source of this element. The accumulated dead bodies and excrement make up **guano**, which is used as a fertilizer. The photograph shows a dock used in the Ballestas Islands (Peru) for loading the guano that the birds on these islands produce.

In living beings, phosphorus is found mainly as part of the DNA, RNA, and ATP in the cells.

Phosphate is present in water in the form of an orthophosphate molecule (PO_4^{-3}).

THE WATER CYCLE

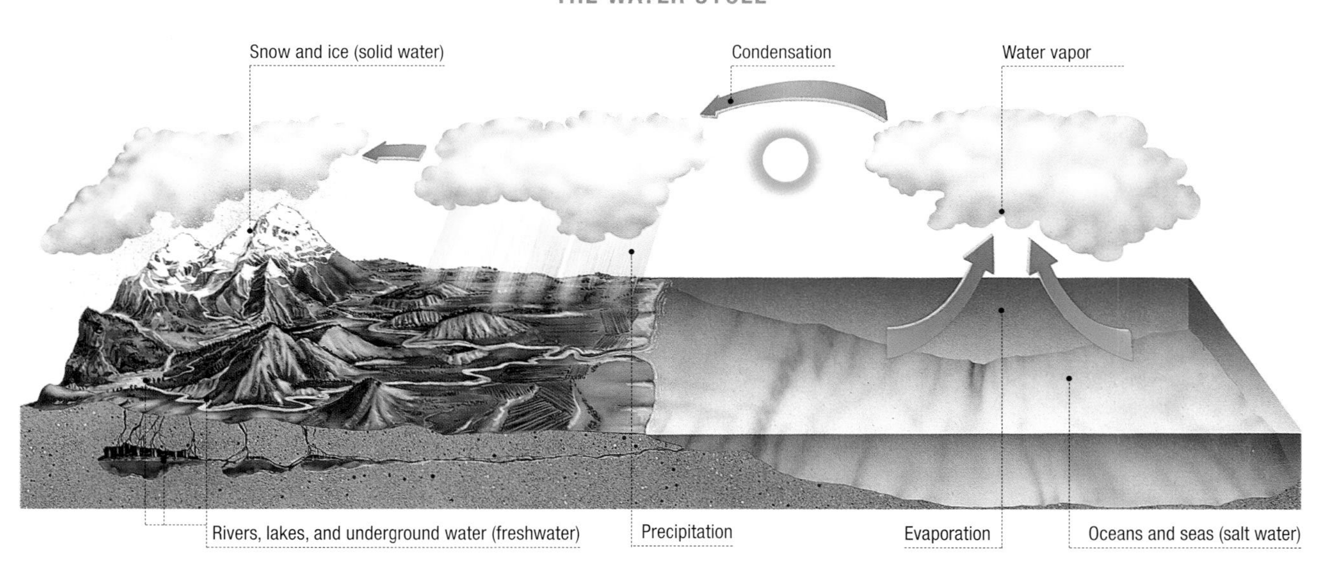

THE WATER CYCLE

Even though **water** is not a chemical element, but instead is a molecule made up of **oxygen** and **hydrogen**, it is essential to all living beings. All chemical reactions that are carried out inside an organism require the presence of water. It is the essential medium for the functioning of **metabolism**, from the production of glucose in plants to the digestion of foods and regulation of temperature in animals. Water too is subject to a constant cycle in nature. It is a very active cycle, both inside and outside organisms.

The body of a mammal is composed of 65% water. In some marine organisms, such as algae and medusas, the proportion may exceed 95%.

Water is a compound of oxygen and hydrogen that can be present in nature in a gaseous state (vapor), a liquid state (flowing water), or a solid state (ice).

THE ECOSYSTEM AND PRODUCTION

Living beings are closely linked to the inert environment in which they develop, especially the **producer** organisms (plants), which extract from it all the materials to manufacture their body structures and the energy necessary to make them function. All other living beings depend entirely on the production of **biomass** by plants. To understand how ecosystems function and to illustrate theories by using numbers, ecologists have to perform a set of generally difficult calculations involving certain parameters such as biomass or **production**.

BIOMASS

Biomass is the total mass of life forms. In a specific **ecosystem** such as a forest, it is the mass of all the living beings that inhabit it. There are several techniques for measuring the biomass of an ecosystem. Generally, they produce fairly approximate values, because to determine them precisely, it would be necessary to gather and to weigh all the organisms in it. This would mean killing many of them, mainly the plants (cutting down the trees, pulling up the grasses, etc.) and that does not make any sense, for if we kill the inhabitants, the ecosystem ceases to exist.

The biomass of this immense Canadian forest consists of animals and especially of the trees that make it up.

MEASURING BIOMASS

Biomass is expressed in units of weight (e.g., lbs, tons/g, kg, metric tons) per unit of area (acre, square foot, ft^2, mi^2, m^2, km^2, etc.).

Production is measured as a unit of weight per unit of time, such as ounces per day or tons per year (g/day or metric tons/year), etc.

The students are doing a study of the biomass of the forest. This is a very labor-intensive job.

CALCULATING THE BIOMASS OF THE FOREST

To calculate the **plant biomass** of a **forest**, we have to mark off a section of ground (e.g., an acre), count the trees, measure the height and the diameter of the trunks, and count the number of branches and their lengths. By using these data (and with a knowledge of the density of the wood), we can come up with the biomass of the wood. To determine the biomass of the leaves, we have to measure the diameter and the height of the crown of the trees. Then we take a typical branch, pluck all the leaves, and put them into a sack. We weigh the contents and then we multiply by the rest of the branches in the area studied. We have to do the same with the vegetation in the **understory**. The last step is to add together the results, and we have the amount of the biomass of the area under study. To calculate the biomass of the whole forest, we have to multiply by the number of acres it occupies.

THE PRODUCTION OF ECOSYSTEMS

Another of the very important parameters in studying ecosystems is the amount of living matter that they produce. Let's use the same example as with the forest. To calculate the plant **production** in one year, we have to determine the quantity of new matter that the trees and the other plants in the **forest** (brush and grasses in the understory) have produced in that period.

BIOMASS PRODUCTION

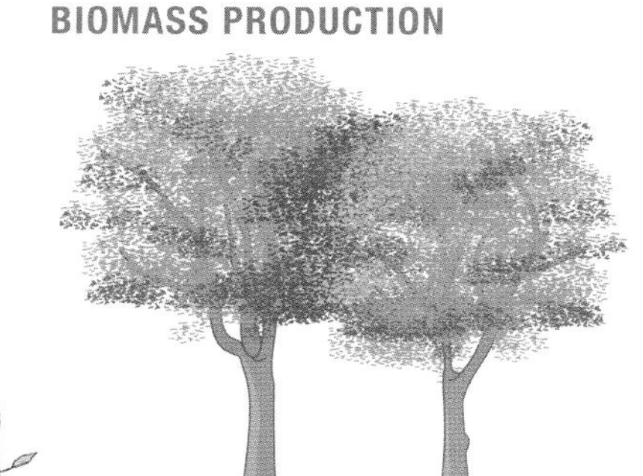

Year 2003: 89 pound/acre (100 kg/ha)

Year 2004: 107 pound/acre (120 kg/ha)

Production = the difference in biomass/time
Production = (107 pound/acre – 100 pounds/acre)/1 year = 7 pounds/acre/year

Productivity is calculated in percentage by dividing the production by the biomass (**productivity** = **production** / **biomass**). For example, if a forest with a biomass of 89 pounds/acre (100 kg/ha) produces 7 pounds/acre (8 kg/ha) per year, its annual productivity is 7%.

In a pine forest, the majority of the biomass corresponds to the trunks and branches of the trees, followed by the roots, and finally, the needles.

PRODUCTION

Production is defined as the increase in biomass per unit of surface area and time.

PRODUCTIVITY IN ECOSYSTEMS

There are a multitude of different **ecosystems** on our planet. All of them maintain a natural balance, which they have reached after functioning for centuries. However, some are more **productive** than others. For example, the **North Pole** is an ecosystem in which populations maintain a perfect balance; however, it is not very productive, because the density of biomass on the ice is very low. On the other hand, the African **savannas** are extremely productive. Each year millions of tons of grass are produced; these are consumed by millions of herbivores, which in turn are prey to thousands of carnivores. The population density in this ecosystem is infinitely greater than that of the frozen Poles. Still, since the entire production of the different **trophic levels** is consumed by the succeeding ones, the biomass of the ecosystems does not increase but remains stable.

Productivity is the relationship established between production and biomass.
Quantitatively, the polar regions are not very productive, but the African savanna is very productive.

USE AND EXPLOITATION OF THE ECOSYSTEM

Even though **ecosystems** achieve a state of equilibrium, they undergo continual change. The different life forms in them produce new **biomass** as they grow, but they also consume other resources, such as the minerals in the soil on the part of the plants, and plant biomass on the part of animals. All these living beings take advantage of the biomass and the resources that the ecosystem offers them in different ways, some as producers, and others as consumers.

MATURE ECOSYSTEMS

Some ecosystems, such as **tropical jungles**, are in a perfect state of equilibrium, and their biomass neither increases nor decreases with time. This means that everything that the plants produce is consumed by the herbivores; in other words, the herbivores consume only what the plants produce. The same is accomplished with the carnivores with respect to the herbivores, and so forth. In the end, production is equal to the loss of biomass because of different factors (consumption, respiration, etc.).

In mature ecosystems, such as a tropical jungle, there is a great variety of species.

SUCCESSION

This is a very important process in **ecosystems**: a series of changes that result over the course of a long time (even centuries) and end up producing a mature ecosystem.

YOUNG ECOSYSTEMS

This term is applied to the ecosystems that exhibit positive **productivity**; in other words, production exceeds consumption. As a result, new **biomass** is created in them as the years go by. This is what happens when a new area is colonized, or, for example when a forest begins to regenerate after it has been destroyed by fire.

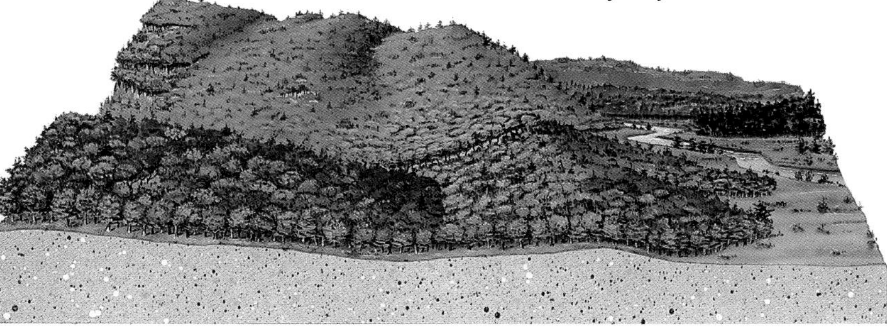

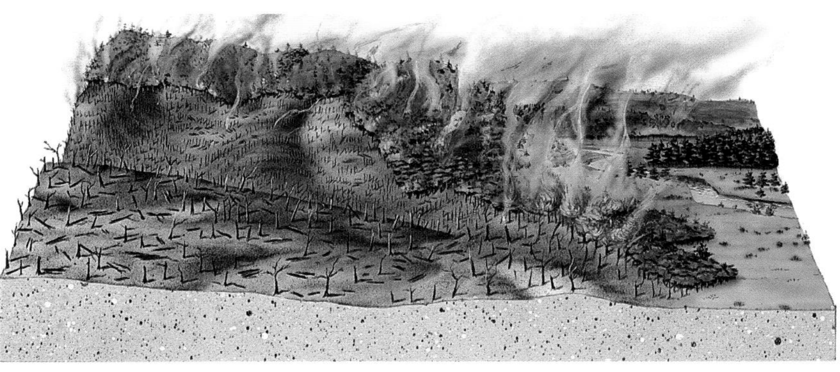

A burned forest is a young ecosystem when new shoots appear.

TROPHIC LEVEL

In studying an ecosystem, it is necessary to classify the life forms and to group them according to how they obtain **matter** and **energy**. Each of these methods is known as a **trophic level**.

AUTOTROPHES

These are living beings that are capable of using solar light for energy and transforming inorganic matter into organic matter for their own body.

THE TROPHIC CHAIN

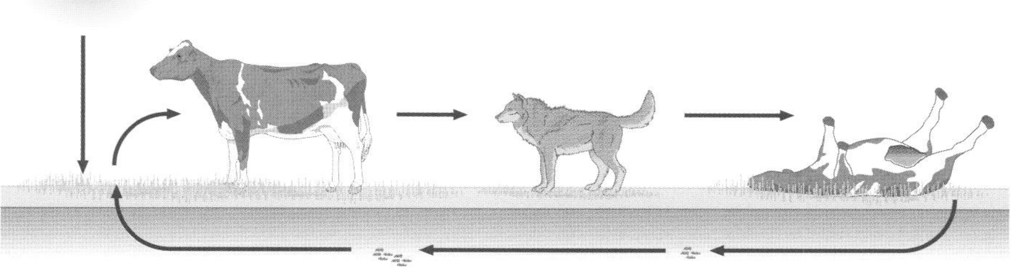

There is another type of autotrophic bacteria that extract energy by breaking the bonds of inorganic molecules. They are referred to as **chemoautotrophs**.

TROPHIC LEVELS

Trophic level	Organism	Type of organism	Energy used	Transformation performed
Producers	Plants	Autotrophs	Solar	Inorganic matter into organic
Primary consumers	Herbivores	Heterotrophs	Chemical	Organic plant matter into animal
Secondary consumers	Carnivores	Heterotrophs	Chemical	Organic animal matter into animal
Decomposers	Bacteria, fungi	Heterotrophs	Chemical	Organic matter into inorganic
Transformers	Bacteria	Autotrophs	Chemical	Inert inorganic matter into mineral fertilizers

In addition to plants (from mosses to trees), other autotrophs are algae and certain bacteria.

HETEROTROPHS

These are living beings that obtain energy from the molecular bonds of organic matter through respiration. These include animals, fungi, protozoa, and a multitude of bacteria.

PRODUCERS

Plants are called **producers** because they are the ones that produce the organic matter in the **ecosystem**. They are closely linked to the physical environment on which they grow, for they use the mineral salts and the water from the soil and the carbon dioxide from the air as raw materials for their own bodies, and solar light for energy. All these components are inorganic, but plants are capable of using them to create organic matter.

Producers commonly are green because they contain chlorophyll, the molecule that makes photosynthesis possible. Still, there are other pigments used in photosynthesis that are not green. That is why we encounter red, orange, and purple plants.

THE ECOLOGICAL PYRAMID

In nature, organisms depend heavily on one another, setting up relationship chains that are so tight that when one species fares poorly another one may be threatened as a result. Plants are capable of fabricating organic matter from minerals, and they form the basis for all the other organisms. These relations that link certain organisms to others are known as **trophic chains**.

ECOLOGICAL NICHE

Within an ecosystem there can be several species that occupy the same trophic level; that is, they feed on the same things. However, each of them is adapted to a series of specific conditions and uses that food in a different way, so that they really do not compete with one another; as a result they can coexist without difficulty. This specialization within the ecosystem is called an ecological niche.

Whales (left) feed on krill that they filter through their strainers; orcas (right) capture fish, birds, and medium-sized mammals. Both of them occupy the same ecological niche, but they don't compete with one another for food.

TROPHIC CHAINS AND NETWORKS ON LAND

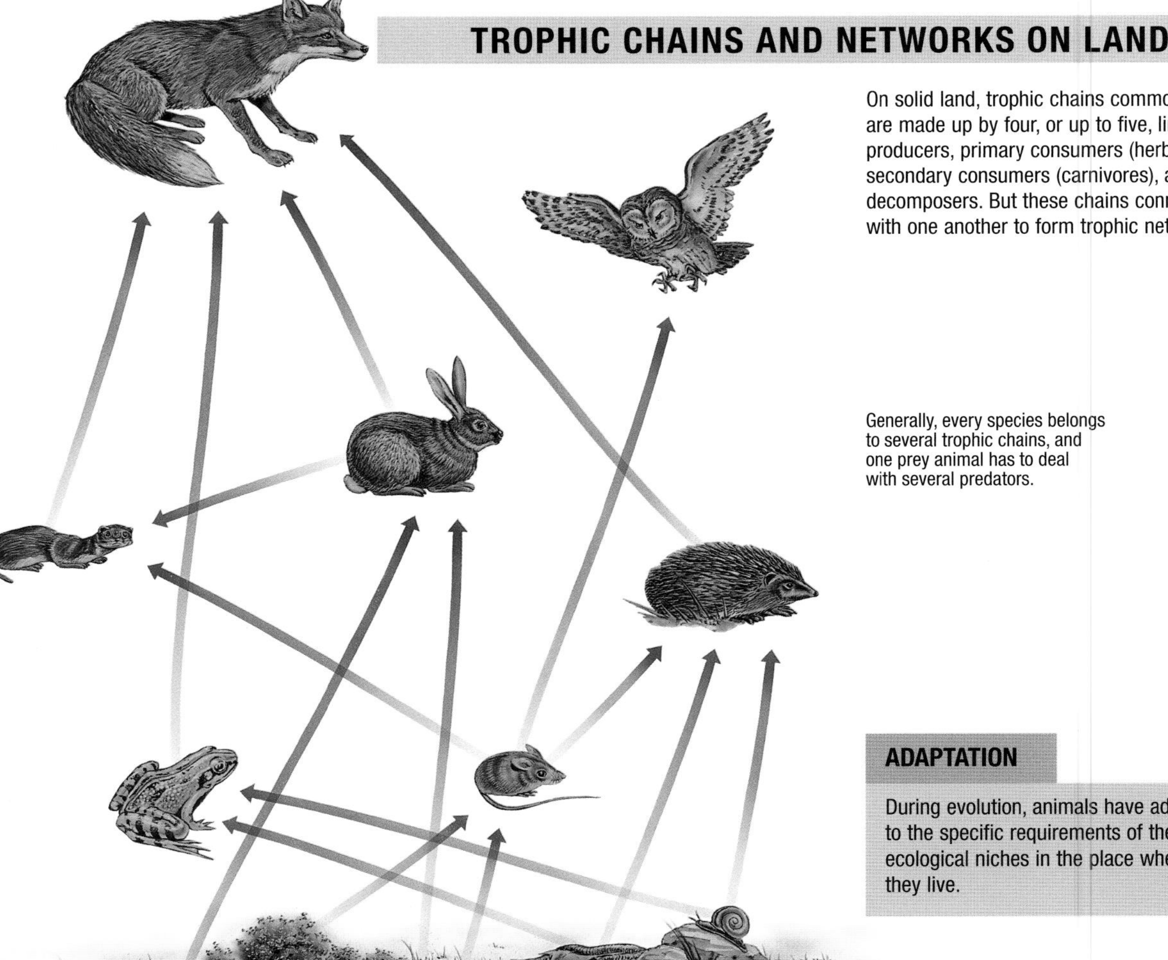

On solid land, trophic chains commonly are made up by four, or up to five, links: producers, primary consumers (herbivores), secondary consumers (carnivores), and decomposers. But these chains connect with one another to form trophic networks.

Generally, every species belongs to several trophic chains, and one prey animal has to deal with several predators.

ADAPTATION

During evolution, animals have adapted to the specific requirements of the ecological niches in the place where they live.

TROPHIC RELATIONSHIPS IN THE OCEAN

In the ocean, the trophic chains are quite a bit different from the ones on land. Their main feature is that they are much longer, since there are tertiary and quaternary consumers, and more. In addition, even though there are medium and large algae, the main mass of producers in the ocean consists of microscopic algae that make up the phytoplankton. Many invertebrates, and in some cases even large vertebrates such as whales, feed on phytoplankton.

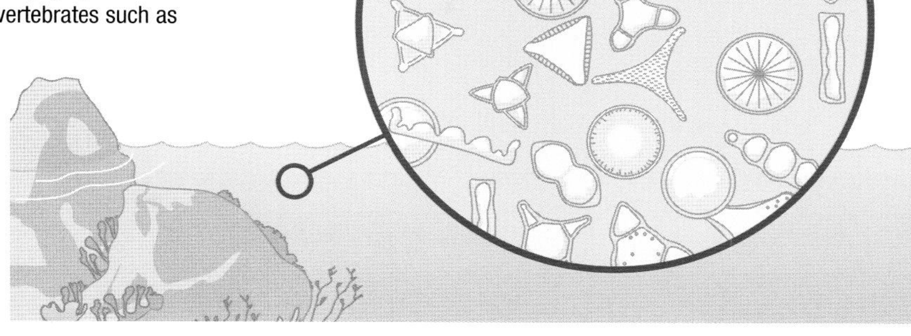

The greatest producing mass in the ocean is made up by phytoplankton, which consists of microscopic algae.

WEIGHT PROPORTIONS AMONG LEVELS IN AN ECOLOGICAL PYRAMID

Trophic level	Examples	Units of weight/acre
Carnivores	Shrew Fox	1
Herbivores	Rabbit Partridge	50
Producers	Apple tree Wheat Blackberries	400

MARINE DECOMPOSERS

These are found on the bottom, because inert organic material tends to sink. They abound on the bottom of the deep sea, where all the organic remains from the upper layer of the ocean end up.

THE TROPHIC OR ECOLOGICAL PYRAMID

Here is a schematic representation of the quantity of biomass that exists in every one of the chains in the trophic networks. Every level is narrower than the one below; for example, there is a greater quantity of plant biomass than animals that eat plants, because otherwise these animals would die of starvation. The same thing occurs with the remaining levels.

EXPLOSIVE GROWTH

Some organisms reproduce very quickly, and, for a certain time, a higher stage on the trophic pyramid may be wider than the one below—for example, the marine zooplankton.

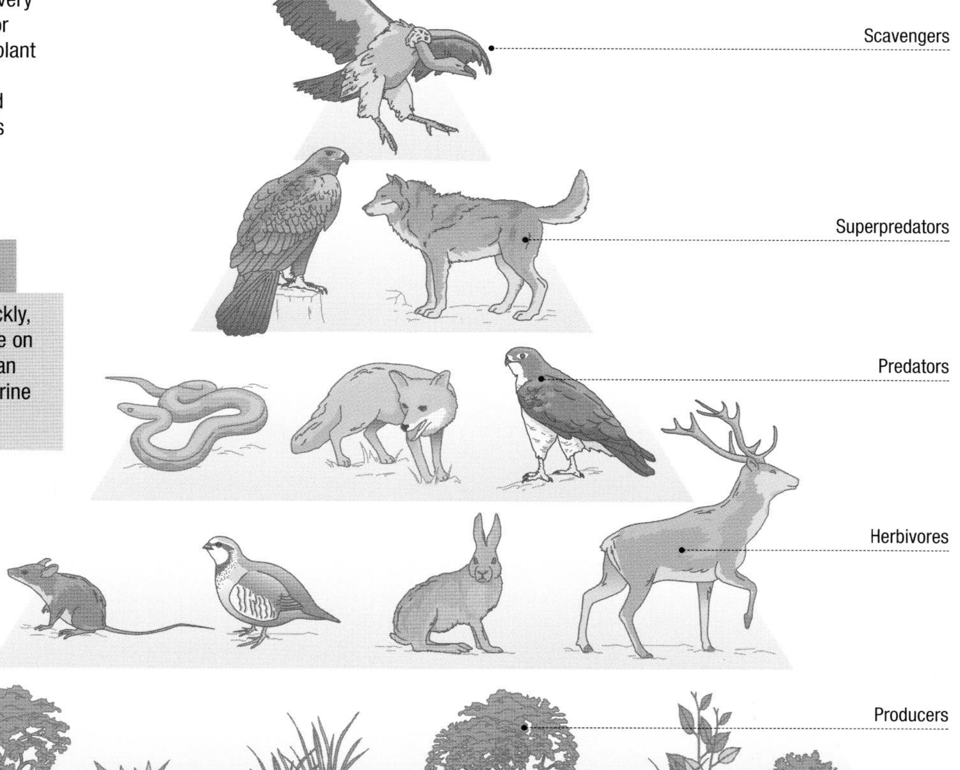

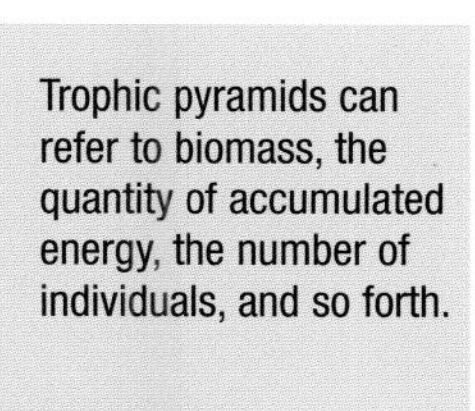

Trophic pyramids can refer to biomass, the quantity of accumulated energy, the number of individuals, and so forth.

COMPETITION AND PREDATION

An ecosystem has many resources of all kinds, and all living beings want them. This gives rise to competition, that is, the struggle of some against others to get what they want. In addition, each living being has its own way of obtaining food.

Plants make food for themselves by using water, minerals from the soil, and light from the Sun. Animals cannot do that, so they either eat plants or they eat other animals. The latter are the predators or hunters.

COMPETING FOR FOOD

All living beings need food to live. For trees, this food is the nutrients that come from the soil and water, so they attempt to get it through deeper or more extensive roots than their neighbors. The first one to arrive will grow more and become stronger. The vultures of the African savanna that find the remains of a zebra fight among themselves to get a piece of meat.

The longer the tree's roots, the more resources it captures (water, minerals, etc.).

Vultures on the African savanna competing to get a piece of giraffe flesh.

When several vultures surge onto a dead body, the most aggressive ones–the hungriest ones–are the first to eat.

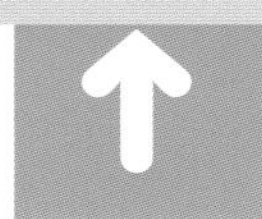

SHARING RESOURCES

Many nocturnal birds of prey, such as owls, and some diurnal ones, such as buzzards, eat rodents, but they do not compete because some hunt at night, and others by day.

When different species of birds occupy a tree to make their nests, some choose the trunk; others, the crown; others, the lower branches, and so forth.

COMPETITON FOR SPACE

In tropical jungles, the crowns of the trees form a type of canopy that keeps the light from reaching the ground. As a result, many plants climb up others to reach higher regions and to get the light they need. This is what many species of climbing plants and vines do. Animals also compete for space, and many of them defend their territory from intruders. A carnivore's territory is all the space it needs to find enough prey to live on.

PREDATORS

Many animals feed on other animals, which they hunt; that is, they are carnivores. They are called predators. Many of them have developed strong claws and teeth or beaks for hunting. The hunting method varies according to species. Thus, wolves hunt deer by chasing them in packs and taking turns in the **pursuit**; the leopard hunts gazelles by racing after them; falcons hunt doves by **diving** from a great height; jaguars hunt peccaries by lying in wait for them in the undergrowth.

Logically, predators have to be faster, stronger, and smarter than their victims to survive. Here are the powerful claws of the leopard; below is the wolf, which commonly hunts in a pack.

NATURE'S CONTROL SYSTEM

The populations of prey animals control the populations of predators, and vice versa.

SUPERPREDATORS

These are hunting animals that are able to feed on other hunters. An example is the golden eagle, which can hunt foxes.

Rabbits are food for many birds of prey. An illness that ravaged the rabbits caused a decline in the number of birds of prey.

Among the mammals, the tiniest hunter is the shrew, which measures a scant two inches (5 cm) and weighs about 0.2 ounce (6 g).

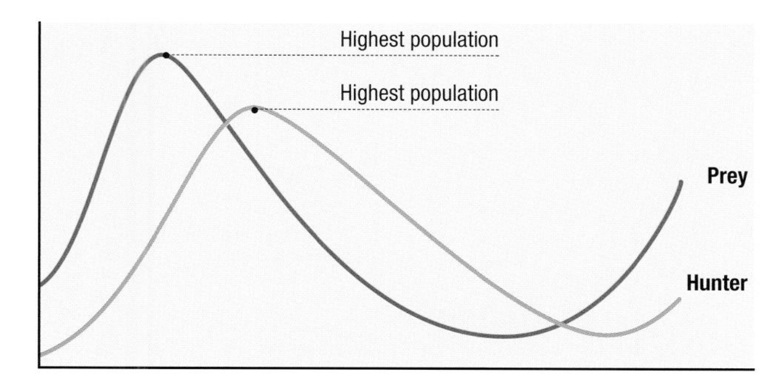

THE PREDATOR–PREY RELATIONSHIP

The greatest number of hunters results when the prey is already on the decline, because they manage to survive a little longer.

In ecosystems there is always a balance between the hunters and their prey. If the number of prey animals increases, the hunters multiply because they have more food; but if there are lots of hunters, they end up killing almost all the prey animals, and most of them will die of starvation. The few surviving prey animals then begin to reproduce, and their population will increase because there are not many hunters. Later on, because the hunters have abundant food, they will again reproduce, and the situation will repeat itself.

The increase in wild boars throughout Europe is caused by the near total disappearance of its principal enemy, the wolf.

POPULATION AND ITS CHANGES

The living beings that inhabit our planet do not appear in isolation; rather, they accumulate in certain places to make up what we refer to as populations. These units are very important in ecology because they can be used to determine how the ecosystem is functioning. However, populations are not permanent but are subject to the changes that living beings experience; in other words, they may increase or decrease. In addition to these numerical changes in their populations, living beings also experience regular changes in location known as migrations, which affect many other populations.

THE INHABITANTS OF THE PLANET

The population is made up of life forms that have some common characteristics; they may be of very different types, depending on what we consider to be common. So, for example, we may speak of the human population of the Earth if we are referring only to the human species. Within a forest there is a population of mammals that includes martens, boars, dormice, mice, and so forth, but we can also speak of the animal population and include mammals, birds, reptiles, insects, and others to distinguish it from the plant population made up of such things as grasses, bushes, and trees.

Expanding populations have a greater number of individuals in lower age classes (young specimens) than in the higher age classes (old individuals).

HISTOGRAM OF THE POPULATION OF EUROPE

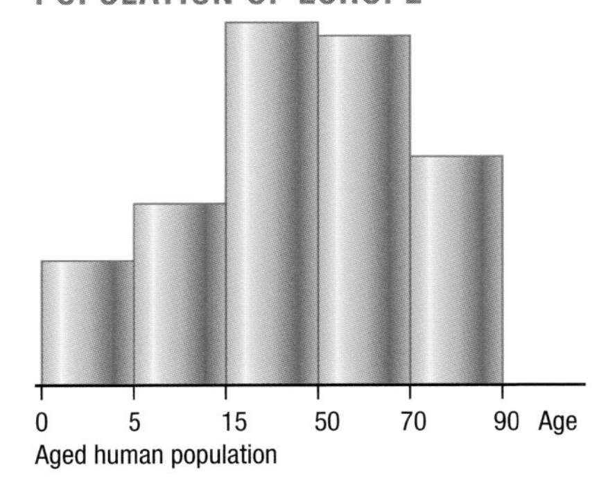

HISTOGRAM OF THE POPULATION OF AFRICA

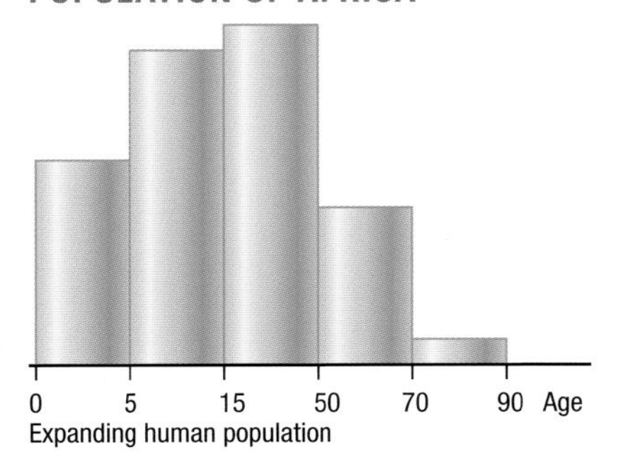

HUMAN POPULATION

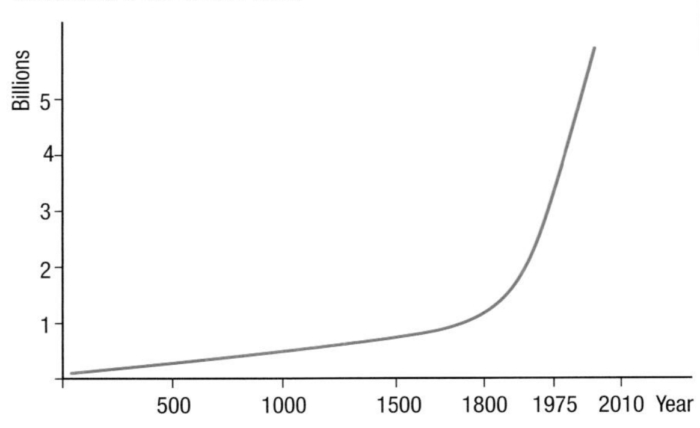

BISON POPULATION

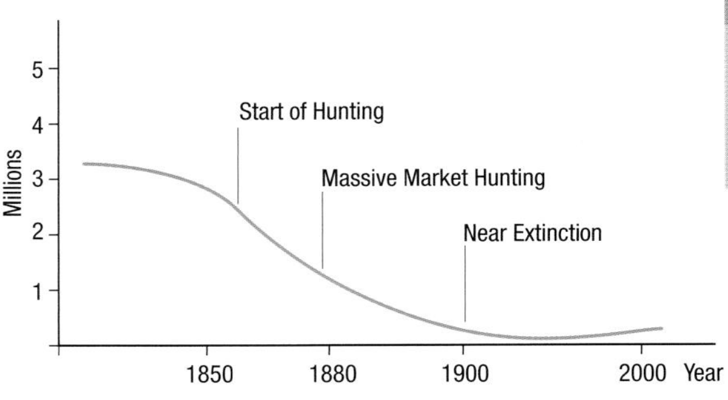

POPULATION STRUCTURE

A method of combining different components of the population, e.g., by age (children, youth, adults, old people).

POPULATION

A population is a set of individuals of the same species (or group) that occupy a given area.

POPULATION CENSUS

Individual birds are banded so that the population of certain species can be tracked. The photograph shows banding a falcon.

In the human species, modern censuses are carried out by filling out a form for each resident of a country. For many animals, individuals are counted and marked; that makes it possible to follow their movements and to count the number of individuals in each group. With birds, bands are usually attached to their legs. With some large mammals, such as bears, radio-transmitter collars are used. With fish, a plastic strip is attached to the base of the dorsal fin. All these marks contain a reference that makes it possible to assign data.

POPULATION CHANGES

Natural populations experience cyclical changes because of the environmental conditions where they live; these can be of a very different magnitude, depending on the species, but they tend to fluctuate around certain mean values. Depending on the amount of available food, the population will have a greater or a lesser number of individuals, and the same occurs with weather conditions. Thus, more weak individuals die in the winter and the population decreases, but, with the arrival of spring, new individuals are born, and the population increases again. At other times, the changes are caused by unusual factors, such as a disease or a plague.

Rodents frequently experience great population fluctuations. Some years they are rare, and in others they become a scourge.

FLUCTUATION

This is a cyclical change in the number of individuals in a population over a certain period of time.

Animals with a low birth rate (e.g., whales) experience few fluctuations in population.

MIGRATIONS

Many animal species undertake regular, long migratory voyages, nearly always because of their need to search for food. Examples include the **gnus** of the African savanna and the reindeer in boreal regions. The best-known migrators are the birds; they nest in high latitudes during the summer when there is plenty of food, and they spend the winters in lower latitudes, where they find the food that is missing in the nesting areas. Many fish also migrate (tuna and salmon, for example), and so do some insects, such as butterflies.

Salmon travel to the headwaters of the rivers to find the right conditions for hatching their eggs.

The Arctic tern travels thousands of miles every year from the North Pole to the South Pole, and back again.

THE RICHNESS OF ECOSYSTEMS

The **life forms** that are present in every **ecosystem** depend on the physical conditions of the environment, including the geology and other factors, but especially the climate. All of that is a prime determinant of the opportunities for plant life and makes it possible for animal life to develop. The ecosystem thus evolves to an optimal state known as **climax**. This is the greatest possible diversity and richness for the environment in question.

CLIMAX

We have already seen in previous chapters that ecosystems are not rigid and static, but rather that they evolve over time. This evolution is what we call **succession**, and the result is a final ecosystem that remains stable, as long as the climate and the physical conditions of the environment likewise avoid major variations. In this condition we say that the ecosystem is **mature**, and the species that make it up evolve slowly but without changing the whole. The input and the outlay of energy and matter are kept continually stable.

PRIMARY SUCCESSION

This is the succession that occurs naturally in an ecosystem.

In the landscape reproduced in the illustration, we can see the distinct stages through which a region passes where the **climax** is a **forest**. First the bedrock is colonized. When the soil forms, grass grows; then low shrubs appear. Later these are replaced by bushes. In a later stage trees grow among the bushes until at last the trees take up the whole area and constitute a forest.

The final ecosystem resulting from a succession is called the climax; it contains the greatest possible biological development for the environmental conditions.

Every region of the planet has a characteristic climax. On the **tundra** it is grass and lichens, since conditions do not allow the growth of larger plants. In the Amazon Basin the climax is the **tropical rain forest**.

SECONDARY SUCCESSION

This is the term used to describe the succession that occurs in an altered ecosystem (e.g., because of a fire) to recover its original condition.

AN EXAMPLE OF DESERT ISLAND COLONIZATION

	Year 0	After 25 years	After 50 years	After 65 years
Lower plants	1	12	61	71
Higher Plants	0	100	140	220
Insects	0	150	500	750
Reptiles	0	2	3	4
Birds	0	15	35	45
Mammals	0	0	3	4

BIOLOGICAL DIVERSITY

In praising the natural riches of any part of the world, we speak of its great biological diversity. This means that it possesses a great number of **different species**. This is a characteristic of the most highly organized ecosystems, in other words, the most stable ones. When an ecosystem arrives at the end of its natural evolution, it is made up of many more species than at the beginning of that transformation, because as it evolves, new **niches** are created; as we have already noted, these are the special places that each species can exploit.

An **immature ecosystem** commonly has one dominant species that is represented in far greater numbers than the other species.

Tropical jungles are characterized by their great biological richness, which is expressed in the form of diversity. There can be hundreds of different species on a single tree.

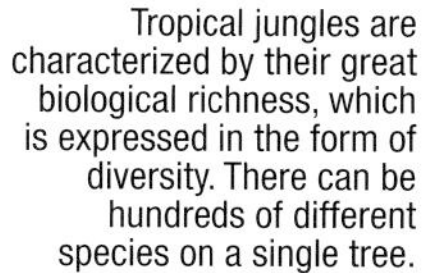

DOMINANT SPECIES

This is the term applied to the species that has a far greater number of individuals than the others in the ecosystem.

DIVERSITY

An ecosystem has great biological diversity if, instead of one dominant species with millions of individuals, there are many species with much smaller populations.

Biological diversity is a guarantee for the life of our planet.

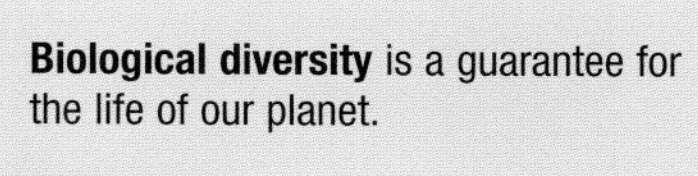

Monocultures, in which only one species is cultivated, are very fragile, for if an epidemic should strike, the entire crop could be wiped out.

The great areas of grain cultivation are highly degraded ecosystems in which the grain (wheat, corn, etc.) is the dominant species. Humans strive to annihilate almost all other species.

SEAS AND COASTS

Nearly three quarters of our planet is covered by the water of the **oceans** and **seas**; in addition, the ocean is where life **first** appeared on Earth. All this makes the ocean a very important environment, even though many of its features are not well known. Conditions for life in the water are different from the ones on solid ground and, as a result, marine organisms display some very distinctive characteristics. The oceans are not a uniform environment; rather, they contain very different regions and zones.

THE MARINE ENVIRONMENT

In the **water** there is an upward thrust that partly compensates for the **force of gravity**, and this means that the force necessary for moving an animal's body is less than on solid ground. As a result, the largest living animals, the **whales**, live in the ocean. Also marine arthropods, such as the **crustaceans**, reach sizes far superior to land arthropods, such as the insects. One further condition that the marine environment nearly always imposes is the need to be equipped with **gills** for taking in oxygen dissolved in the water.

Lobster

Temperature contrasts are much smaller in the ocean than on solid ground. The ocean cools off and heats up less, and more slowly, than the air.

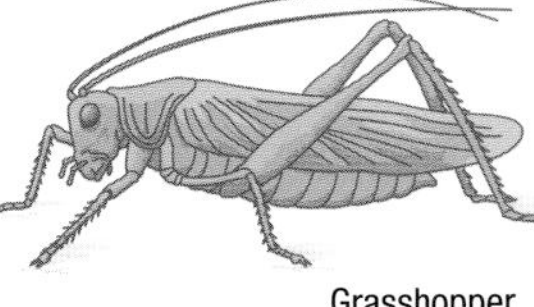

Crustaceans, which are marine arthropods, commonly are larger than insects, which are land arthropods.

Grasshopper

SALINITY

This refers to the number of grams of salt per liter of seawater. The average salinity of the ocean is 35 parts per thousand; in other words, there are 35 grams of salt in each liter of water.

PRINCIPAL COMPONENTS OF SEAWATER (g/l)

Component	g/l
Sodium chloride ($NaCl$)	26.51
Magnesium chloride ($MgCl_2$)	2.25
Magnesium sulfate ($MgSO_4$)	3.30
Calcium chloride ($CaCl_2$)	1.14
Potassium chloride (KCl)	0.72

OCEAN SURFACE AREAS

Ocean	Sq. mi.	% of total
Pacific	64,169,000	46.0
Atlantic	33,411,000	23.9
Indian	28,343,000	20.3
Arctic	3,661,000	2.6
All Seas	10,139,000	7.2

THE REGIONS OF THE SEA

THE SEA REGIONS

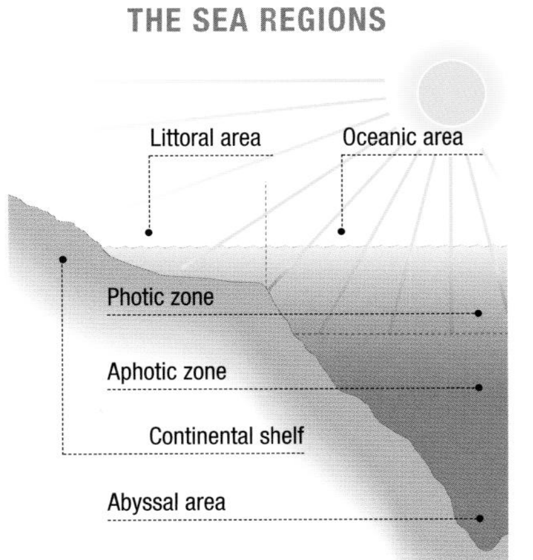

Down to a depth of 500–650 feet (150–200 meters)

Down to a depth of 36,000 feet (11,000 meters)

If we look at the sea from the shore, we see a uniform surface but that is the case only on the outside. Underneath there are areas of different depths, which go by different names. Also, the areas closest to the coast, which are called littoral regions, are different from the areas of open sea. One important division is the one that designates the depth to which light can penetrate. Marine plants can live above this limit, but there is a shadowy world that begins at this level, beyond the sun's light, and where the **deep-sea** areas reach great depths.

THE APHOTIC ZONE

Zone beyond the reach of the sun's rays. It is the deepest part of the ocean, in some places more than 36,000 ft (11,000 m) deep.

THE PHOTIC ZONE

This is the zone where the sun's rays penetrate. Its depth depends on the transparency of the water, but it generally does not exceed 650 feet (200 meters).

COASTS

These are the borders between the sea and terra firma. It is an area of great **biological activity**, since there are plants and animals that live in both environments; also, there are others that generally live in one of them but make use of the other, for example to get food. There are many types of coasts, from level, **sandy** ones to coasts made up by rocky inlets and cliffs. The conditions for life in a coastal zone are very hard for both plants and animals. This is an extremely varied **ecosystem** with great **biological productivity** and **diversity**.

Most of the species that are fished for human consumption live on the continental shelf.

The continental shelf is the relatively narrow area that surrounds the continents and reaches a depth of 650 feet (200 meters).

THE FIVE LARGEST SEAS

Name	Surface Area (sq. mile)	Maximum Depth (feet)
Coral Sea	1,849,814	30,000
Arabian Sea	1,422,014	19,000
South China Sea	1,148,499	15,000
Caribbean Sea	971,394	25,000
Mediterranean	969,116	16,000

SEA ANIMALS AND PLANTS

Dolphins

The organisms that inhabit the sea can live **free** on the bottom, as the sea urchins and lobsters do, or they may remain attached to rocks, such as mussels and dark seaweed. Others live far from the bottom. Some are very small and **float**, as do many protozoa, and others, such as fish and whales, are quite large and swim actively.

Many marine organisms, such as mussels, seaweed, and others, live attached to rocks.

PLANKTON

This term encompasses all the very tiny organisms that live floating in the water.

BENTHIC AND PELAGIC ORGANISMS

Benthic organisms are the ones that live at various depths on the bottom of the sea. Pelagic organisms are ones that live swimming freely in the water, both in littoral regions and in the open sea.

RIVERS AND LAKES

From the point at which the first droplets of water rise from a spring or a snowfield to where a large river empties into the sea, there is a tremendous variety of different environmental and ecological conditions. As a result, the flora and fauna in each of these zones are different.

A river is a constantly changing ecosystem, and it has tremendous importance for the life around it. Lakes are the bodies of water that look like small seas inside the continents and share some characteristics with seas.

RIVERS

Sometimes rivers are justifiably compared with the veins of the natural world, because the **water** they carry performs a function similar to that of the **blood** that flows in our bodies. Just as the veins reach the cells to feed them with **nutrients** and **oxygen**, rivers and their tributaries carry the water that is necessary for life. Vegetation can grow near them, and many animals come to them to drink. In addition, their waters contain a rich assortment of fauna and flora.

BASIN

A basin is all the watercourses that end up in the same river. The different basins are separated by large geographical features (such as mountains and deserts).

Rivers are excellent indicators of the condition of the ecosystems around them.

THE PARTS OF A RIVER

Course	Conditions	Bottom	Vegetation	Fauna
Upper	Strong current, low temperature	Rocks and boulders	Very scant or nonexistent	Good swimmers or attached to bottom
Middle	Moderate current, medium temperature	Stones and sand	Clustered in pools	Medium swimmers
Lower	Slow current, high temperature	Sand and mud	Very abundant	Poor swimmers or floaters

The Amazon is the world's longest river. It is 4,356 miles (7,025 km) long and flows west to east in South America.

THE PARTS OF A RIVER

Upper course

Middle course

Lower course

THE PARTS OF A RIVER AND ITS INHABITANTS

A river can be divided into parts, depending on its physical structure or its inhabitants; the latter depend on the physical conditions that the river offers at every stretch. At the headwaters, the river has little flow, but its grade may be very steep, and the animals and plants that live in it have to adapt to the cold temperatures, a very strong current, and a bottom that commonly is made up of large rocks. These conditions moderate as the river travels toward the end, its mouth. Here the rate of flow is at its greatest, but the current is much reduced, and the main features of the substrate are mud and sand.

FLOW RATE

This is the quantity of water that flows in a river. It is measured in units of volume (e.g., cubic feet) per unit of time (e.g., seconds). At its mouth, the Amazon has an average flow rate of 7,000,000 cubic feet per second.

LAKES

These are fairly modern **ecosystems**, for they have been around for only a short time, based on a planetary time scale. These are bodies of quiet water or ones that have minimal current when crossed by a river, and they occupy natural depressions in the terrain. Some lakes occur in old volcanic **craters** that have been filled by rainwater; others occur in sinkholes in the terrain or in depressions left by **glaciers**. In some ways lakes are similar to seas, especially in their division into separate zones based on depth. Lakes also fulfill a very important ecological role, especially in arid regions.

LAKE TITICACA

Although Lake Titicaca is located at an altitude of about 12,506 feet (3,812 m) in an arid area of the Andean high plains, its great surface area of over 3,219 square miles has an influence on the climate of the surrounding region, making it possible to grow crops species that do not exist elsewhere on the high plains.

Lakes, especially large ones, lessen the climatic effects of the areas where they are located.

FRESHWATER LIFE
(Some typical species from different places around the globe)

Location	Vegetation	Invertebrates	Vertebrates
River: upper course	Trees on the bank	Crabs, some insects	Kingfisher, trout
River: middle course	Trees on bank, reedbeds	Crabs, bivalves, insect larvae	Barbels, piranhas, otters, frogs, turtles, crocodiles, hippopotamuses, ducks
River: lower course	Trees on bank, reedbeds, submerged plants	Bivalves, worms, insect larvae	Eels, electric fish, plaice, river dolphins, herons
Prairie lake	Trees on bank, reed-beds, submerged plants	Insect larvae, worms	Barbels, tilapias, ducks, herons

The largest lake in the world is the Caspian Sea, with an area of 143,206 square miles; it is a saltwater lake.

FRESH AND DEEP

The largest freshwater lake in the world is Lake Superior, with an area of 31,691 square miles. The deepest lake in the world is Lake Baikal in Siberia, which reaches a depth of over 5,000 feet.

LAKES: THEIR ZONES AND INHABITANTS

Similarly to what happens in a sea, there are several zones in a lake from the shore to the bottom. There is a scale of vegetation from the trees that grow on solid ground or partly submerged in certain cases, to the plants that root on the bottom and can live completely submerged. In each of these zones there are different types of animals. There are species that are adapted to areas of stagnant water or minimal current and which can be found in lakes but not rivers.

LIFE IN A POND

FORESTS AND JUNGLES

Forest ecosystems are much more than a mere collection of trees. They are communities in which plants and animals live together in perfect balance. Depending on the regions of the globe where they occur, these forest ecosystems come in very diverse forms, from the **taiga** of the cold regions to the tropical **rain forests** and **jungles**.

TAIGA

This is a forest transformation that makes up an extensive border region in the high latitudes in the **northern hemisphere**, with very low winter temperatures and a relatively short summer. **Conifers** (firs and pines) predominate almost exclusively, and deciduous trees are found only on the shores of lakes. Except for a few animals that remain year-round in the taiga, most of them **migrate** in the fall to lower latitudes.

Moose

The animals that live in the taiga are adapted to the harshness of the climate.

The taiga reaches its greatest extent in Siberia, with a length of 2,930 miles and a width of 610 miles.

INHABITANTS OF THE TAIGA

Some of the most representative permanent inhabitants of these forests are eagle owls, moose, Siberian tigers, wolverines, pine martens, squirrels, goshawks, and others.

Goshawk

THE TEMPERATE FOREST

The temperate forest extends throughout the **temperate regions** in both hemispheres, with four clearly defined seasons. It is also called a **deciduous forest**, because the dominant trees are the flat leaf variety (deciduous), such as beech, birch, linden, oak, hazelnut, elm, and others. The **underbrush** is very lush (strawberries, blueberries, heather, lilies, and others). This results in an abundance of plant food, which allows the existence of rich and varied wildlife (deer, bear, foxes, wolves, roe deer, badgers, dormice, grouse, eagles, kites, frogs, salamanders, ants, butterflies, and so forth).

Grouse

The American deciduous forest has more diversity than its Eurasian counterpart, because the mountains in the Americas are oriented north and south, and that makes it easier for the species to migrate.

The beech forests of Spain's Ordesa Valley are a good example of a temperate forest.

THE MEDITERRANEAN FOREST

This is a forest made up of trees with coriaceous (leathery) leaves that stay on all year long (holm oak, cork, olive) and are resistant to the summer **drought**. This forest is typical of the Mediterranean region, but it also exists in South Africa and in middle latitudes of the eastern coasts of North and South America. Some of the inhabitants are boar, foxes, rabbits, lynx, and vultures.

The cork tree, typical of the Mediterranean forest differs from the holm oak in its bark, which is thick and light; it is the material from which corks are made.

OPEN FOREST

This is an open holm-oak forest with alternating meadows and trees. It is of human origin, but it is a very important ecosystem for wild animals.

Boar

Rabbit

THE AMAZON RAIN FOREST

This is a forest of **broadleaf** evergreen trees, with high environmental humidity, and which is home to one of the greatest concentrations of species on the planet. It forms an enormous, low canopy under which a uniform **microclimate** is maintained throughout the year. It contains many species of valuable woods, plus ferns, orchids, and climbing plants. Snakes, toucans, hummingbirds, parrots, piranhas, caimans, jaguars, tapirs, anacondas, and many creatures make up its animal population.

Toucan

THE FRAGILITY OF THE ECOSYSTEM

The **tropical rain forests** are the main lungs of the Earth, but they are also ecosystems that are very sensitive to changes, and any alteration can destroy them irreversibly.

The Amazon rain forest occupies an area of about 2.7 million square miles.

THE MONSOON FOREST

A large part of the monsoon forest has disappeared through wood cutting and forest fires.

This resembles the Amazon rain forest, but it grows in Southeast Asia. The climate of this region is not uniform; there is a distinction between a **wet season** (with the **monsoon**) and a dry season. Thus, the vegetation also adapts to withstand these changes. The inhabitants of this region include snakes, tigers, leopards, panthers, pheasants, gibbons, orangutans, rhinoceroses, and others.

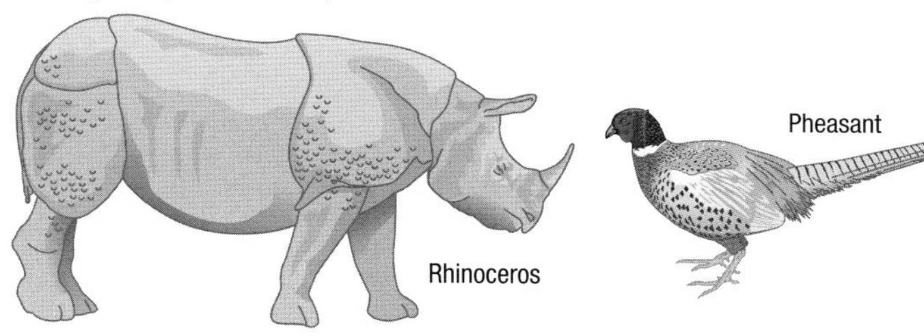

MANGROVE SWAMPS

Mangrove swamps are a very special type of ecosystem made up of a forest that grows on the water in **tropical** and **subtropical coastal** areas. The trees that make it up and give it its name are **mangroves**, which have aquatic roots that allow them to gain ground on the sea. This is an environment where there is a combination of marine life (fish, crabs, etc.) and land life (e.g., monkeys and birds).

Crab

FISH WITH LUNGS

These fish are capable of breathing atmospheric air, which allows them to move around in the branches of trees in mangrove swamps and on damp ground.

The mangrove swamp is host to a multitude of mainly amphibious fauna.

THE GREAT GRASSY PLAINS

One type of very extensive plant formation is the one made up by the grasses. These grasses reach different heights, depending on the climate, and, on flat terrain, they produce very special ecosystems such as the African savannas, the South American **prairies** and **pampas**, the Eurasian **steppes**, and the North American **plains**. This is an ideal world for the large herbivores and the winged hunters. Sight is an important sense in this environment, and speed in running is an important quality for hunting and avoiding becoming the hunted.

THE AFRICAN SAVANNA

This is one of the most well-known and popular landscapes. There are two types of savanna, one with **grasses** and the other with **bushes**, where there are great expanses of underbrush. In both cases, there are also large, isolated trees or small clusters of them. The climate is characterized by high average temperatures and the alternation of two seasons, a **dry season** and a **rainy** one. This means that when the pastures get used up, many of the herbivores (such as the gnus) have to **migrate** to other areas.

Giraffe

Water buffalo

The characteristic trees of the African savanna are acacias and the giant baobabs.

Total rainfall in the savanna is about 273 inches (70 cm) per year.

The African savanna during the dry season.

TYPICAL INHABITANTS OF THE SAVANNA

Hunters	Lions, cheetahs, leopards, hyenas, wild dogs, eagles
Prey	Elephants, giraffes, gazelles, gnus, antelopes, zebras, ostriches, buffalo, monkeys, baboons

THE SOUTH AMERICAN PRAIRIES AND PAMPAS

The Argentine pampa, with the southern foothills of the Andes in the background.

In the plains, the average annual temperature varies between 77°F and 82°F (25–28°C), and there are about 59 inches (1,500 mm) of rain per year.

TYPICAL INHABITANTS OF THE PAMPAS AND PLAINS

Plains	Bats, harpy eagles, peccaries, mule deer, jaguar, anacondas, armadillos, capybaras, foxes, crested screamers, caracaras
Pampas:	Viscachas, owls, hares, ostriches, tinamous, ovenbirds

In the northern part of South America there are great, flat areas where the high average temperatures throughout the year and the rains favor the development of herbaceous vegetation. Here too there are two seasons, a **dry** one and a **rainy** one. At the other end of the continent, in the southern third and to the east of the Andes, are located the **pampas**, which are great prairies with a temperate climate, where low grasses predominate. There are four seasons, and based on precipitation there is a distinction between the **dry pampa** and the **wet pampa**.

Armadillo

Rhea, an American ostrich

In the pampa, the average annual temperature is about 61°F (16°C), and the annual rainfall varies between about 16 inches (400 mm) (dry pampa) and 39 inches (1,000 mm) (wet pampa).

THE EURASIAN STEPPES

The steppes constitute a broad band of land extending from Eastern Europe to the extreme east of Asia. They are located to the south of the **deciduous** forest, and they lead up to the desert regions in central Asia. The climate is **continental** in type, with cold winters and very hot summers. There is an absolute predominance of grasses, and a nearly total absence of trees. Depending on the amount of rainfall, there is a distinction between the high, the mixed, and the low plains. Over the centuries, many areas have been converted for growing grain.

Jackal

Woodchuck

TYPICAL INHABITANTS OF THE STEPPES

Hunters	Wolves, jackals, eagles
Prey	Hamsters, susliks, mice, wood-chucks, saigas, bustards

The Siberian steppes in the Autonomous Republic of Khakassia.

PRECIPITATION AND TEMPERATURE

The Eurasian steppes are about 2,450 miles (4,000 km) long. There the annual precipitation varies between about 12 inches (300 mm) and 39 inches (1,000 mm). In the summer the temperature exceeds 86°F (30°C), and in the winter it goes down to –22°F (–30°C).

THE NORTH AMERICAN PLAINS

Bison next to a lake on the North American plains.

Badger

This is a formation very similar to the **Eurasian steppes** but with greater botanical richness. The plains extend from central Canada to northern Mexico. Nowadays they exist in unaltered form only in certain **reserves** and **national parks**, because much of the plains have been turned into pasturelands and croplands for growing grains. However, in contrast to the Eurasian steppes, the transformation has been so rapid and the animal life has not been able to adapt, so much of it has disappeared.

PRECIPITATION AND TEMPERATURE

The annual precipitation on the North American plains varies between about 10 inches (250 mm) and 47 inches (1,200 mm). In the summer the temperatures exceed 86°F (30°C), and in the winter they get down to –22°F (–30°C).

TYPICAL INHABITANTS OF THE PLAINS

Hunters	Wolves, jackals, eagles
Prey	bison, prairie dogs, pronghorn, prairie hens, badgers

Wolf

ARID ENVIRONMENTS

This is the term used to designate all **ecosystems** in which one main characteristic is a lack of **water**. Thus, the clearest example of this type of environment is the **desert**, both cold and hot. However, we also include others that may contain water, even in abundance, but in a form that **plants** cannot use, so the end result is the same. These types of environments include the **tundra** and the **polar regions**.

HOT DESERTS

These are tremendous extensions of land nearly or totally devoid of vegetation, where the average **temperatures** are very high; during the night, however, they may drop significantly. These regions also have very scant **precipitation**, which often falls only every so many years. This reduces the plant life to just a few plants that are able to sprout and grow during the few rainy days, or ones that grow in **oases**.

OASES

Oases are points in the desert located over an aquifer and where the water reaches the surface, allowing the growth of abundant vegetation.

In deserts, such as this one in Libya, there are very few manifestations of life.

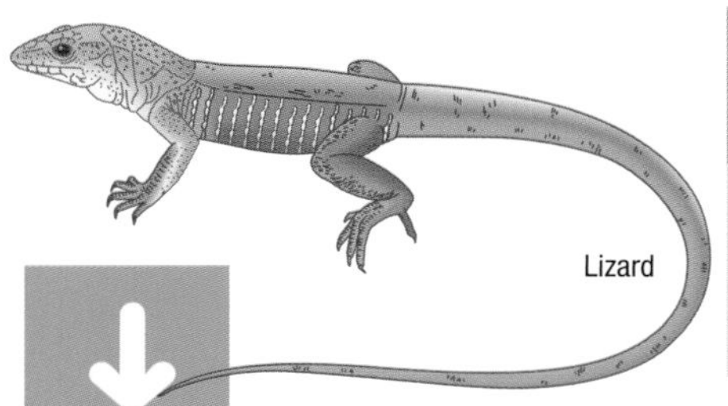

Lizard

The Sahara Desert covers an area of 3,512,600 square miles.

North American deserts are characterized by their tremendous varieties of cacti, including some gigantic ones, such as the saguaros.

SOME TYPICAL DESERT INHABITANTS

African deserts	Dromedaries, oryx, gazelles, lizards, sand grouse, imperial sand grouse, foxes, Chinese desert cats, grasshoppers, lynx, falcons
North American deserts	ground squirrels, Gila monsters, rattlesnakes, roadrunners, red lynx, kit foxes, gerbils

SAHEL

This is a sub-Saharan region measuring about 3,350 miles long by 120 to 360 miles wide located to the south of the Sahara, where semidesert conditions prevail.

Camel caravan on the Khongoryn dunes in Mongolia's Gobi Desert

COLD DESERTS

The main difference between this type of desert and the previous ones is the **average** annual **temperature**, which is much lower. In the winter it can get down to –22°F (–30°C), but the heat in the summer can be torrid. However, the fauna have **adapted** in similar ways, so many animals live underground to protect themselves from the temperature (whether hot or cold), and they come out only in search of food. Others, such as the **camels**, put on a thick winter coat to withstand the cold; they lose it when summer arrives.

The Gobi Desert occupies an area of about 772,000 square miles.

DESERT ANIMALS

The Bactrian camel (with two humps) and the Mongolian gazelle are very typical animals of the central Asian deserts. Others, such as the wolf and the falcon, come from the surrounding steppes.

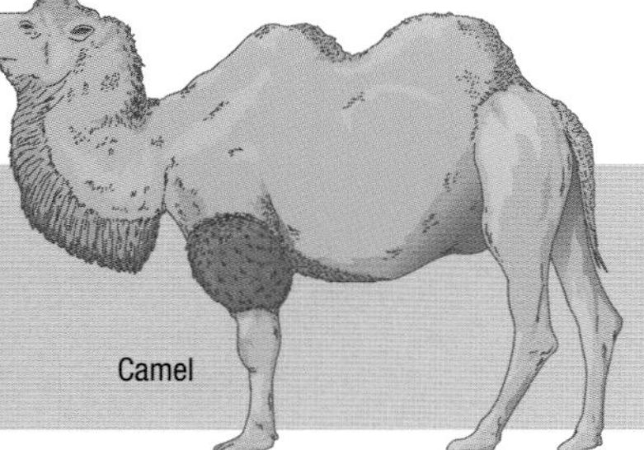

Camel

DUNES

Dunes are piles of sand formed by the wind when the displaced sand piles up against an obstacle.

TUNDRA

The tundra is made up of large expanses of flat terrain where the soil is always frozen starting at a certain depth. When the surface layer thaws during the summer, large marshes are formed. The layer of soil is very thin, so only **mosses**, **lichens**, and certain **grasses** are able to grow. This environment is ideal for **aquatic birds**, which come to nest in this region by the millions. The aquatic plants germinate and produce fruit in a very short time, and the insects likewise complete their cycle very quickly, so there is a great abundance of food.

Mallard duck

The vegetation of the tundra is very poor, but it is very resistant to the harshness of the climate.

PERMAFROST

The part of the soil that remains frozen all year is referred to as permafrost.

The tundra covers the extreme northern area of North America and Eurasia, and there is a small amount of tundra in the southern end of South America.

Antarctica covers an area of 5,558,400 square miles.

THE POLAR REGIONS

The two ends of the Earth, the **poles**, make up the polar regions. During the winter, the night lasts 24 hours a day, and in the summer the daylight lasts for 24 hours. The temperatures are very low, and a small part of the ice melts only in the summer. There is scant precipitation (and even none in some parts of Antarctica), but there is enough to supply the necessary snow. One important difference is that the North Pole, the **Arctic**, is an ice cap that floats on the ocean, whereas the South Pole, the **Antarctic**, is a continent covered with a thick layer of ice.

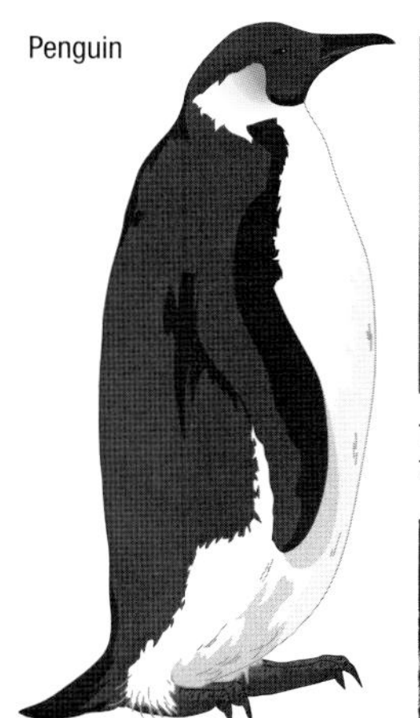

Penguin

The ice of Antarctica covers a continent, in contrast to the Arctic, under whose ice there is no solid ground.

Seagull

ANTARCTIC VEGETATION

In some places that are free of ice, mosses and lichens grow.

The Arctic offers no conditions for establishing human settlements. Inside the Arctic Polar Circle only a few small fishing villages are possible.

TYPICAL FAUNA OF ANTARCTICA

Birds	Penguins, seagulls, terns
Mammals	Seals, elephant seals, leopard seals, whales

TYPICAL FAUNA OF THE ARCTIC

Birds	Seagulls, terns
Mammals	Polar bears, seals, walruses, whales

The Arctic Ocean covers an area of about 5,404,000 square miles and reaches a maximum depth of 16,892 feet.

MOUNTAINS AND HIGHLANDS

Beyond a certain altitude, high ecosystems generally require a major effort on the part of life forms, which have to adapt to less oxygen in the air, lower temperatures, and, in the case of mountains, a rugged topography. These challenges have given rise to specialized flora and fauna, but they have also transformed many of these environments into a refuge for many species that are threatened elsewhere by human pressure.

MOUNTAINS

Mountains arise as the result of a **folding** in the Earth's crust. Depending on the location where they appear, the **steepness**, and the **height** they attain, they may completely isolate certain regions of the planet from others. This is one of the main ecological functions of mountains: acting as a **geographic barrier** that hinders the dispersion of many species. Another important function is that they constitute an ecosystem with conditions different from the ones that prevail around them.

There are 14 peaks over 26,250 feet (8,000 m), and all are located in the great mountain range of the Himalayas. There are scarcely seven or eight people in the whole world who have climbed them all.

The summit of Cotopaxi, an active volcano in the Ecuadorian Andes, is 19,498 feet (5,943 m) high; it has a permanent snow cover over 16,400 feet (5,000 m).

A GEOGRAPHICAL BARRIER

This is any insurmountable barrier for a species, which prevents its dispersion. It favors the speciation of large botanical and zoological groups.

THE DISTRIBUTION OF VEGETATION

As you climb up from the valley toward the summit, you encounter different types of **vegetation**. After the fields come the **deciduous forests**, which are then replaced by **conifers**, which are more resistant to the cold. The last level is occupied by meadows separated from the forest by an area of scrubland, because trees cannot grow at that altitude. The upper limit of vegetation is marked by the year-round snows, where sometimes there are also **glaciers**.

THE PRINCIPAL MOUNTAINS OF THE WORLD

Africa	Kilimanjaro (Tanzania)	18,275 feet
North America	Mount McKinley (Alaska)	20,320 feet
Central and South America	Aconcagua (Argentina)	22,834 feet
Asia	Mount Everest (Nepal)	29,028 feet
Europe	Montblanc (France)	15,771 feet
Oceania	Wilhelm (Australia)	13,975 feet

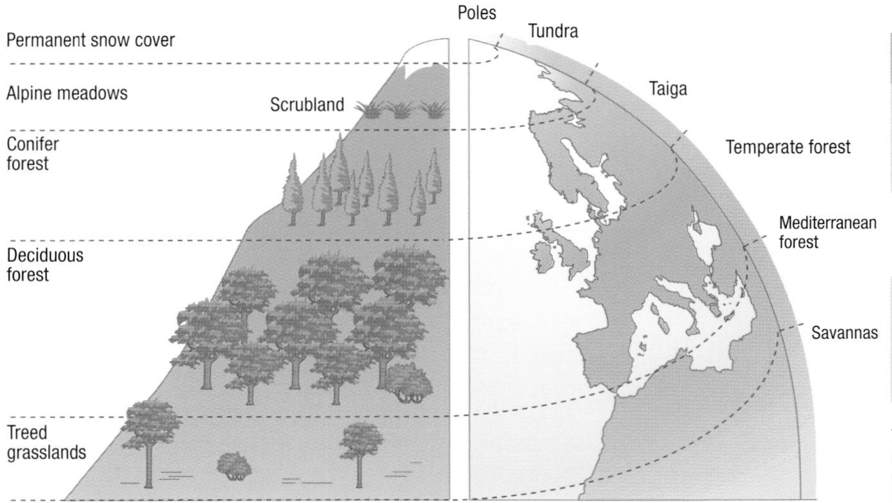

ALTITUDINAL LEVEL

This term refers to each of the strips of vegetation that characterize a certain altitude on a mountain.

With respect to the surrounding area, on a mountain the temperature goes down about one degree centigrade for every 150 meter increase in altitude.

The altitudinal levels are roughly equivalent to the different latitudes of the planet.

TABLELANDS AND HIGH PLAINS

High plateaus and great plains located at high altitude between mountains share the same characteristics as the mountains, such as reduced oxygen and reduced temperature, as well as the conditions found on the plains, that is, the absence of major topographical features. These tend to be areas of scant precipitation, so the vegetation is nearly always of a type similar to that of the **steppes** or the **tundra**.

ANDEAN HIGH PLAINS

The Andean high plains are located at heights of 12,000 to 13,000 feet (3,600 to 4,000 meters). In the north they are called **páramos**, and in the south, **punas**.

Some of the driest areas on the planet, such as the Atacama Desert, are found in the high plains of the Andes. This desert covers more than 50,200 square miles at an average altitude of 2,000 feet (600 m).

The immense Lake Titicaca is located at an altitude of over 13,000 feet (4,000 m) on the Andes high plains between Bolivia and Peru.

The dryness of these regions is mainly because of the screening effect produced by the mountain chains that surround them and keep out clouds laden with moisture.

TIBETAN HIGH PLAINS

This is an extensive sloping region covering some 463,300 square miles (1,200,000 sq. km) that runs from about 8,900 feet (2,700 m) at the lowest point to about 16,500 feet (5,000 m) at the highest. As the altitude increases, the vegetation and fauna become increasingly sparse.

In both the mountains and the high plateaus, solar radiation is more intense because of a thinner intervening air layer, and this encourages the appearance of mutations. The photo shows the Tibetan high plains.

SOURCES OF POLLUTION

Polluting an environment means adding some element or material that disrupts the natural functioning of the ecosystem; consequently, the life forms that inhabit it are affected in one way or another. There are three basic causes of pollution in an environment: accumulating material that does not break down, introducing some substance that turns out to be poisonous for life forms living there, and finally, adding too many nutrients.

NONDEGRADABLE AND DISAGREEABLE

When some material fails to integrate into the natural cycle because the living organisms cannot make use of it, we say that it does not degrade. In that way, the material accumulates in the environment and takes up available space. This is what happens in the case of plastics and rubber tires. For example, a plastic bag on the floor of a forest or a meadow keeps any kind of plant from growing beneath it, because it blocks the passage of both light and air.

HEAVY METALS

These are industrial waste products that are toxic to living beings and act like poisons in ecosystems.

When poisons are used to kill rats, we run the risk of having these toxic materials accumulate in such high quantities in the rodent's body that cats, foxes, and owls will also get poisoned when they eat the rats.

In cases where it is necessary to use poisons, we must always choose ones that do not build up in nature.

BIODEGRADABLE

Materials that can be degraded naturally are referred to as biodegradable.

Accumulation of nondegradable waste, especially plastics, may seriously affect forest growth.

Sea turtles confuse plastic bags with the medusas, which they eat, and swallow them. Once inside their digestive track, they cause a blockage and interfere with the passage of other foods. The turtle ends up dying of starvation.

TOXINS

Insecticides, herbicides, and other poisons are chemical products that are used to kill animals, fungi, and plants that bother us in some way; one example is an infestation of aphids in the garden. But the products that kill aphids are also toxic to other living beings in the environment, such as the ladybugs that feed on the aphids and the birds that eat the ladybugs. That is how the poisons get passed from one animal to another in the food chain, affecting more living beings at every turn.

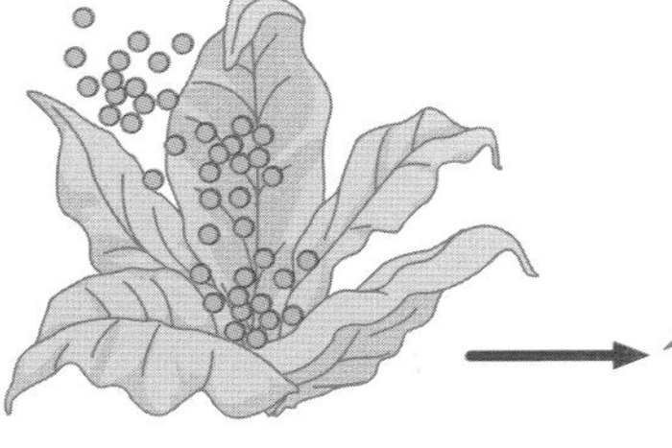

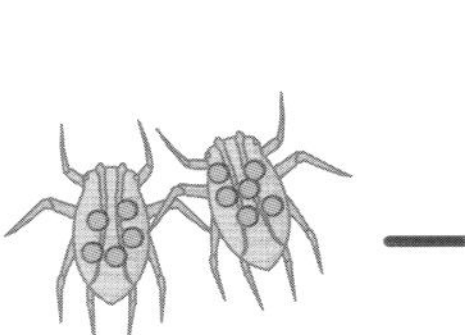

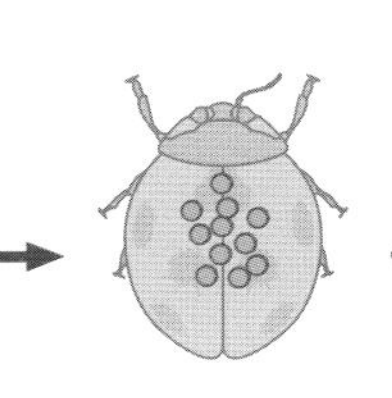

The Toxin Chain

CONTAMINATION THROUGH NUTRIENTS

This is a type of pollution in which the environment is enriched excessively with products that plants exploit for growth. This generally results from an excess of phosphorus and nitrogen, which are components that commonly are used in fertilizers to make fields produce more. If too much nitrogen or phosphorus is added to the natural environment, the balance among organisms is disrupted and the plants grow in excess. In the case of a lake, the algae grow so much that they use up all the oxygen and the animals die.

There are so many nutrients in this lake that there is an excess of algae, which color the water green.

The products that produce the greatest amount of eutrophication are detergents and fertilizers, because they contain a large amount of phosphorus and nitrogen.

EUTROPHICATION

This is the term that is applied to an excess of nutrients in an environment (generally aquatic) that causes excessive plant growth.

EXOTIC SPECIES

The introduction of the rabbit into Australia caused a disaster, because it successfully competed for food with the marsupials.

Although exotic species cannot be considered a source of pollution per se, their introduction into an ecosystem can prove very harmful. In most cases, the organisms are incapable of living in an alien environment; however, on occasion some species is capable of adapting, and then it occupies the same ecological niche as the native species, which clearly suffer as a consequence. When this animal is a predator, often the native species have no defense mechanisms against it.

When an exotic animal is let loose (e.g., a bird) in an inappropriate place, it may cause serious harm to nature.

Sometimes fishermen introduce fish that have sporting value, and the result is the disappearance of the fish that are native to the area.

FIELD ECOLOGY

Ecology is one of the most complex sciences from both theoretical and practical viewpoints, for it is multidisciplinary; in other words, it encompasses physics, chemistry, zoology, botany, meteorology, and many other scientific fields. There are many techniques that are used to obtain information on ecosystems, their productivity, the relationships among inhabitants, explanations on the behavior of the species, and so forth. Some of these techniques are as simple as mere observation, but, in other cases, complicated measurements must be made.

SOME INDISPENSABLE TOOLS

There are a series of parameters that must always be kept in mind when studying ecosystems. First of all we have to situate in space the ecosystem on which we are working. To that end we need at least a map, and if possible, a compass. Another fairly important instrument is a thermometer, since temperature is one of the most influential parameters in the behavior of living beings. The list of tools that can be used in studying nature is endless.

Field work is one of the most appealing and interesting kinds od study. A well-directed and well-equipped group can make very interesting verifications and discoveries.

It is necessary to use precise instruments to obtain data to validate scientific studies. The photograph shows people watching aquatic birds from a camouflaged blind.

In national parks, nature reserves, and so forth, it is against regulations to touch or pick up minerals, plants, and animals. The photograph shows Porcelain Basin in Yellowstone National Park in Wyoming.

Instrument	Purpose
Binoculars	Looking at animals and landscapes
Compass	Determining location in space
Altimeter	Measuring altitude
Measuring tape	Performing length measurements
Boxes or bags	Collecting samples
Magnifying glass	Observing details
Map	Determining geographical location, altitude, latitude, slope of the terrain, distances, etc.
Knife or other cutting implement	Opening up fruits, small animals, etc.
Watch, chronometer	Determining position in time, figuring times
Field guides	Classifying features of the ecosystem (life forms, stones, types of soil, etc.)
Thermometer	Determining the temperature
Barometer	Determining atmospheric pressure
Microscope	Observing microorganisms
Notebook and pencil	Taking notes, drawing, etc.
Camera	Taking photos
Computer	Data storage

EXPERIMENTATION

It is really difficult to conduct experiments in ecology, because ecosystems are very large and they depend on so many things that it is extremely difficult to reproduce natural conditions in a laboratory in a controlled fashion. Thus, every practice that is carried out to study some feature of the ecosystem can be defined as an invention. The results have to be subjected to complicated mathematical and statistical calculations, even though there are also some simple ones that make it possible to reproduce the phenomena in nature.

An attempt has been made in this experiment to create an ecosystem isolated from the outside world and controlling the environmental conditions of several different environments (forest, desert, etc.)

PRESERVING NATURE

Whenever you study nature you must avoid destroying it. You must never pick protected plants or take endangered animals, even if you are not in a protected area.

PLANT RESPIRATION

One of the most classical experiments in ecology (in the branch of plant physiology) involves showing that plants generate **oxygen** during **photosynthesis**. There are several ways to do this. One of the most common ones is to place an aquatic plant inside an aquarium and to cover it with an upside-down glass without any air. Leave it in the light for a whole day. You will see that a bubble containing oxygen has formed. Repeat the same experiment, but this time leave it in the dark. No air bubble forms because photosynthesis does not take place.

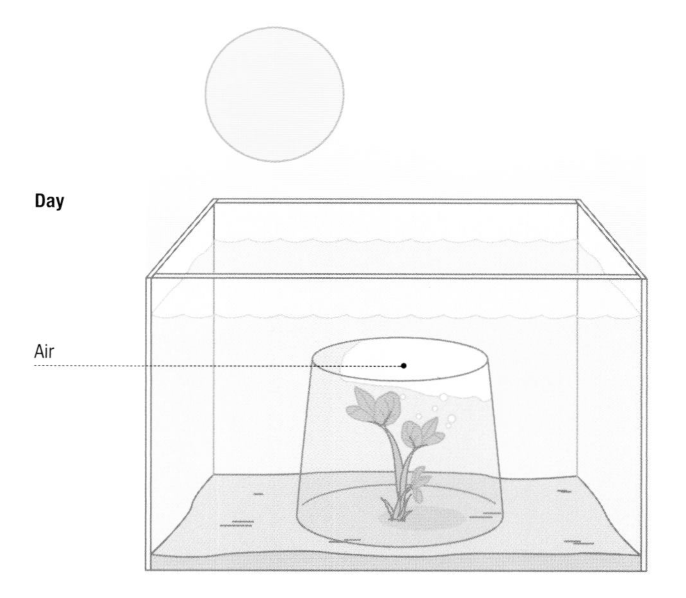

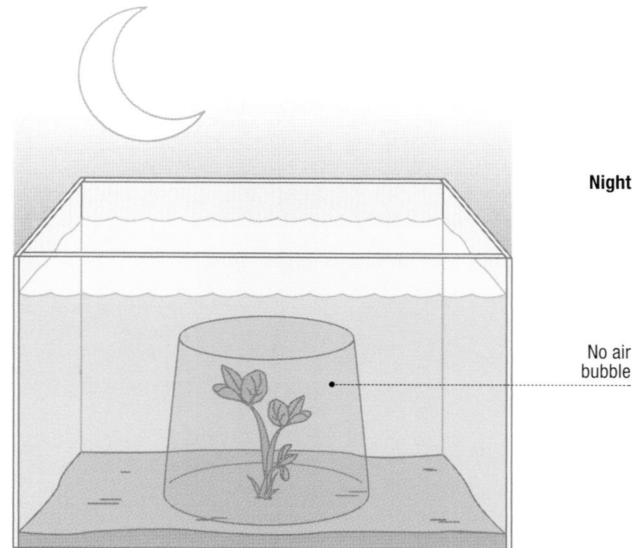

In the demonstrated experiment you can measure the volume of air created by the plant: multiply the area of the base of the glass by the depth of the bubble.

Calculate the same values three times in a row. You will discover small differences in the results. The more carefully you perform the measurements, the closer the results will be to one another.

MAKING AN HERBARIUM

One basis for studying ecosystems involves drawing up a census of the life forms that inhabit them. An herbarium assembles a list of plants from a specific area, and this is one good way to start getting a thorough knowledge of the environment that surrounds us, because it requires that we get out into it many times and that always is much more instructive than getting information from a book.

You have to go out into the field with a large folder and a good number of pages from a newspaper. When you find a plant that interests you (it is a good idea to bring along a guide for identifying them), you have to pull it up by the roots or cut off a piece (in the case of large plants) and place it between the pieces of paper, along with notes about what species it is, and when, where, and under what conditions it was collected. Back home you have to keep it for a few days with a weight on top; a couple of books will work. When the plant is finally dry, you can put it into a photo album with all the explanations.

Some herbariums can become real works of art. Botanical gardens exhibit some very important herbariums dating from the eighteenth and nineteenth centuries that were used to establish the systems of botanical nomenclature.

Collecting specimens for the herbarium.

STUDYING FAUNA

When studying fauna, it is preferable to make a field folder with drawings and photographs of the various species in the area, because it would be a shame to kill them and dry them for storage in a display case.

If you are attracted to a protected plant or you are in a protected area, it is better (and mandatory) to substitute a photograph for the physical plant.

INDIRECT TRACES

In field studies, many times, the researcher has to trust signs that show the presence of a certain creature without seeing it directly. For example, if you want to conduct a census of animals that are difficult to locate for any reason, it is important to recognize the traces they leave in the environment, such as droppings, nests, den openings, feathers, tracks, leftover food, and so forth. Keep in mind that your presence frightens many animals, and they will try to keep you from seeing them.

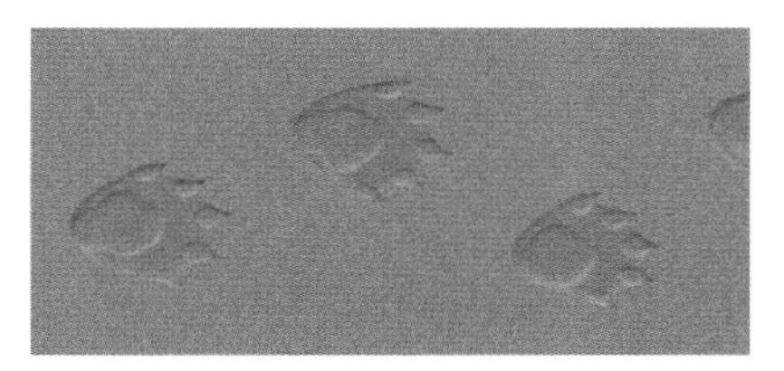

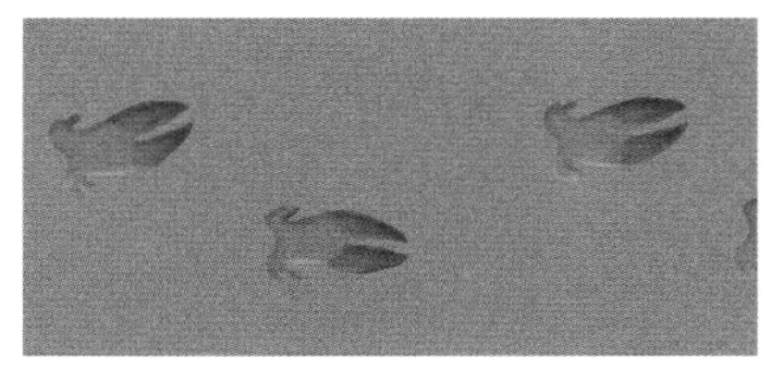

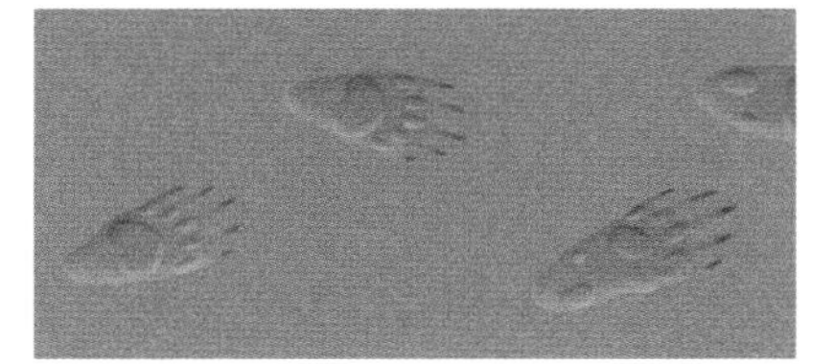

Tracks are one of the best types of evidence confirming the presence of living beings in the environment. Studying tracks can even reveal characteristics of fossil organisms. The only drawback is that this is a complicated discipline, and there are not many people who can really identify the animals that belong to a specific set of tracks.

On the left, a badger den; on the right, droppings from the same animal.

In studying predatory birds, we can identify their food by collecting and analyzing their pellets or castings. These are a type of ball that they regurgitate after eating their prey; they are made up of the indigestible parts of the animal, such as the fur and bones.

OCEANOGRAPHIC VESSELS

There are laboratory ships for studying marine environments; these ships are equipped with very specialized machines, instruments, and tools for research. Scientists in various disciplines travel on these ships; they must work in conjunction with one another to assure the quality of the research. Generally the principal mission of these expeditions is to gather samples that later are analyzed in universities and marine research centers.

Marine research vessel

JACQUES COUSTEAU

Jacques Cousteau (1910–1997) was one of the greatest marine researchers of all times. A professional engineer, he invented numerous devices for accessing the bottom of the oceans (scuba diving and others), and, along with his team, he performed countless dives in which new features of this inaccessible environment were discovered.

The *Hesperides* is a Spanish marine research vessel that has made several expeditions to Antarctica.

WHY IS MATHEMATICS IMPORTANT IN ECOLOGY?

Science is based on quantifiable data. For example, in physics it is known that water freezes at 32°F (0°C) and boils at 212°F (100°C); in chemistry it is known that an oxygen atom combines with two hydrogen atoms to make up one molecule of water, and so forth. However, in the study of populations, there is no certainty that they will remain static: they may remain in balance, increase, or decrease depending on conditions. Nor do we know with certainty when a hurricane will develop, how strong it will be, and where it will go. However, there is a repetitive pattern in these phenomena. It makes it possible to study the same parameters time and again and to make certain predictions and a model of how an ecosystem will function.

Many natural phenomena, such as hurricanes, can be predicted, but how they will act is much harder to determine.

VOLTERRA'S MODEL

The Italian physicist Vito Volterra studied the behavior of predator and prey populations in ecosystems, and he believed in a numerical model that shows that when the population of herbivores expands, after a certain time, that of the carnivores does likewise.

Currently there are many types of modern technology that can be applied to the study of nature. One of the most advanced ones is tracking the land surface covered by plants by means of satellites.

ECOLOGICAL AGRICULTURE

The course of history has taken us from a nonaggressive use of our environment to its destruction. Agriculture has participated very aggressively in this process. The first farmers scarcely caused any damage to the natural environment, but, as the human population and its needs for food increased, the impact of farming activity became more severe, to the point that it now includes the use of substances that are poisonous to many organisms. Now we are trying to return to nonaggressive methods like the one known as ecological agriculture.

PESTICIDES

Until the nineteenth century, agriculture caused practically no damage to the environment. It is true that people occupied ground that was robbed from nature, but the techniques used were not very aggressive, and, in most cases, they were even beneficial for the area's biodiversity. However, starting in 1950, pesticides came into wider use. At first people thought this was a great way to avoid the tremendous annual crop losses caused by insect infestations, fungi, and weeds. But with time it became evident that the nature in the areas where these phytosanitary products were applied was quickly being depleted. Currently, agricultural research groups are working to come up with techniques for controlling insect plagues without affecting the environment.

The abuse of pesticides has produced some harmful animals that are immune to them and has harmed many other animals.

Pesticides accumulate in land ecosystems just as they do in freshwater ecosystems. In this case the toxin is incorporated into the phytophagous insects (ones that eat plants); in turn these fall prey to predatory insects and the birds that eat all these insects.

The concentration of pesticides is so great at the end of the food chain that it can affect the health of animals, prevent them from reproducing correctly, and even cause death.

BIOACCUMULATION OF TOXINS

The pesticides that are applied to farm fields can be carried away by rainwater and can travel to other places for which they were not intended, generally freshwater ecosystems (where most of them dissolve). There they become part of the food chain through the plankton, and they keep accumulating in increasing quantities in the bodies of these organisms. The fish that eat this plankton build up a quantity that amounts to the sum of all that is contained in their food. The same thing happens with the animals that eat the fish (which add to their bodies all the accumulated poison in each fish they eat). Thus, the quantity of pesticide keeps getting more concentrated in every link of the chain.

OUR HEALTH

We humans are not immune to the harmful effects of pesticides, because we too are animals that are part of the food chains.

PESTICIDE CONCENTRATION

Water	0.010 mg/L
Plankton	3.6 mg/kg
Planktonic fish	7.2 mg/kg
Predatory fish	157 mg/kg
Fish-eating birds and mammals	1,780 mg/kg

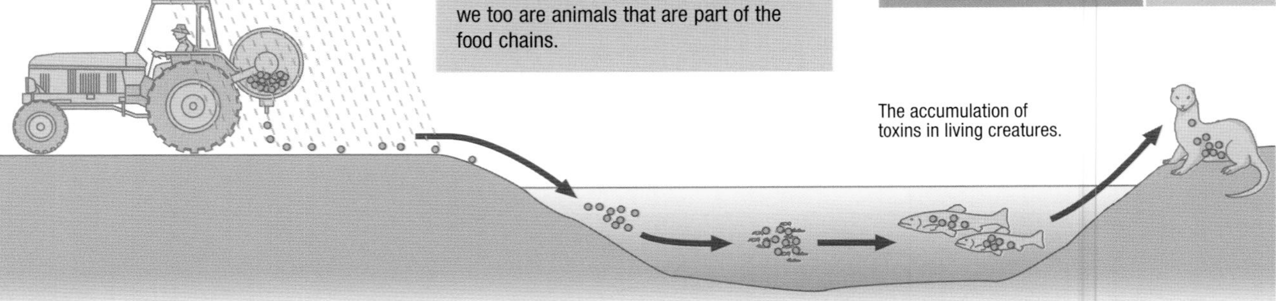

The accumulation of toxins in living creatures.

BIOLOGICAL AGRICULTURE

Nature has resources for controlling the populations of different species. It is very rare for any of them to create an excess and harm the others. This principle can be applied to crop fields, which really are nothing more than one more type of ecosystem. The insects that feed on the plants find such a rich environment that they are able to reproduce in great numbers and with no problems, until they constitute a plague. However, these insects have natural enemies, which in most cases are predatory insects. If we increase the presence of predators, we can keep the population of phytophagous insects from ballooning and thus avoid an infestation.

Nowadays chemical means are used to combat plagues of aphids, but this can be dangerous to other animals.

Ladybugs eat aphids and scarcely have any effect on the plants. If we increase the numbers of ladybugs, it can be said that we are using a natural or ecological pesticide.

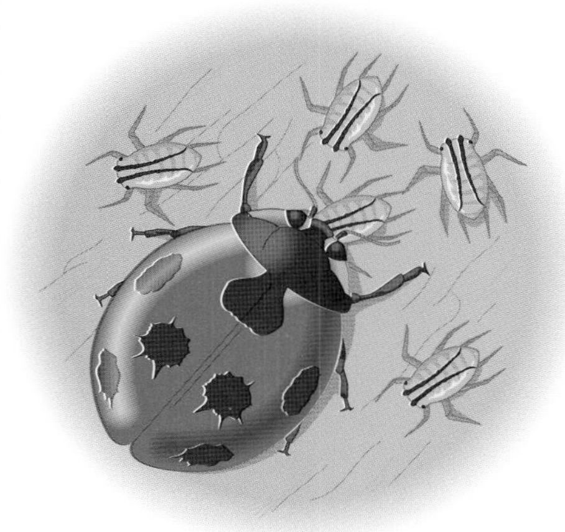

INTEGRATED AGRICULTURE

This combines modern and traditional techniques. To compensate for the lower productivity of biological agriculture, it makes use of pesticides only in extreme cases and in the smallest possible quantities. The presence of pests and their predators is under continuous study, to avoid harming the latter. This makes it possible to avoid massive use of these toxic substances.

The fertilizer used must be the necessary and appropriate one. When there is too much, it is diluted with rain or irrigation water and enters the soil, and from there it travels to the aquifers, which become contaminated.

POPULATION CONTROL METHODS USED IN ECOLOGICAL AGRICULTURE

Natural control	Introducing predators of pest organisms
Ecosystem creation	Allowing the growth of a fringe area around the crop field so that natural predators can live
Crop rotation	Changing the type of crop in each season so that conditions are never right for pests
Crops adapted to local conditions	Using local (natural) varieties that are more resistant to pests

BEES, THE FRUITGROWER'S ALLY

Bees are used in agriculture to maximize the pollination of fruit trees and to increase fruit production. When a bee lands on a flower to get the nectar, it becomes covered with a multitude of pollen grains that stick to its body. When it gets to a different flower, one of these grains of pollen may get introduced into the female part of the flower and fertilize it. Then this flower will turn into a fruit. This natural pollination mechanism can be maximized by placing beehives in fruit orchards. The more bees, the greater the reproductive success of the trees (and thus, their production).

The proper use of irrigation water is an essential method in ecological agriculture, because water is a very scarce commodity in most countries of the world.

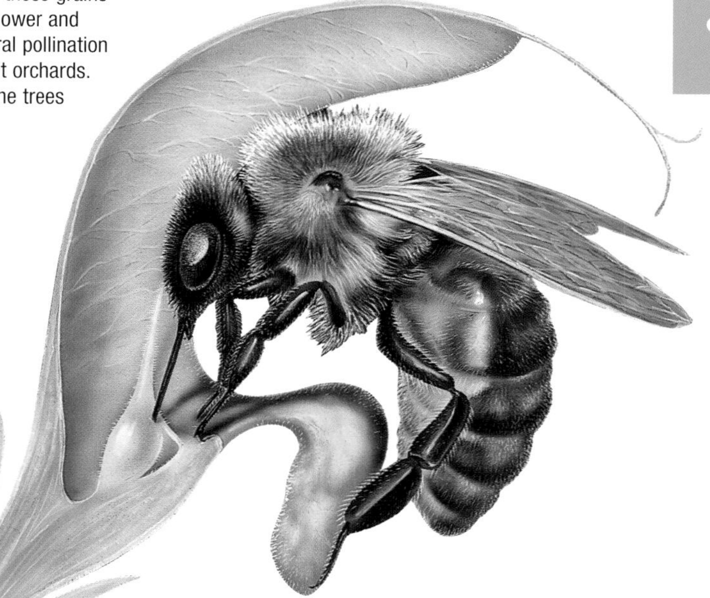

Pollination by bees leads to a greater harvest. That is why beehives (in the photograph) are placed near orchards.

TRANSPORTATION

The advance of automobile technology and the new social habits produced during the preceding century have greatly increased people's mobility worldwide. The automobiles that use the highways every day use gasoline or diesel as a fuel, and they generate a tremendous amount of carbon monoxide and dioxide, plus other compounds, that pollute the atmosphere and warm the air.

TRANSPORTATION IN LARGE CITIES

Urban centers, and especially large cities, are the hottest spots with respect to transportation. These are locations with a high population density, to which is added the multitude of people who travel there for work or in search of services. Many of these people who travel from one point to another in the city or who come in from elsewhere do so in a vehicle, and the most common one is the automobile. All the **emissions** from the exhaust pipes of these automobiles, plus the basic structure of a city (tall buildings that interfere with the proper ventilation of the streets), contribute to a cloud that envelops the urban center, which experiences elevated temperatures and a high concentration of toxic by-products.

Noise is one of the main problems caused by city traffic. Sound pollution is significant, easily reaching 80 decibels in an area of heavy traffic.

The pollution and the congestion caused by the massive use of the automobile are a good argument for using public transportation.

MINIMIZING THE USE OF PRIVATE VEHICLES

In many industrialized countries, the occupancy rate of automobiles is just 1.3 people per vehicle. To avoid this, carpooling is being promoted to reduce costs and pollution.

THE MAIN POLLUTANTS PRODUCED BY GASOLINES AND DIESEL FUEL

Carbon monoxide (CO)
Carbon dioxide (CO_2)
Nitrogen oxide (NO)
Unburned hydrocarbons
Lead

PUBLIC TRANSPORTATION

There is a far more rational and ecological alternative for getting around in a city: the use of public transportation. Currently, in nearly all cities there is a network of internal transportation made up of buses, and in large metropolitan areas there are also subways and streetcars. These high-capacity vehicles can carry many people, and that greatly reduces the consumption of energy per passenger. In addition, the average speed commonly is far greater than that of an automobile; subways travel on an exclusive route (which means there are no traffic jams) and buses commonly have their own lane, which makes it easier for them to get around.

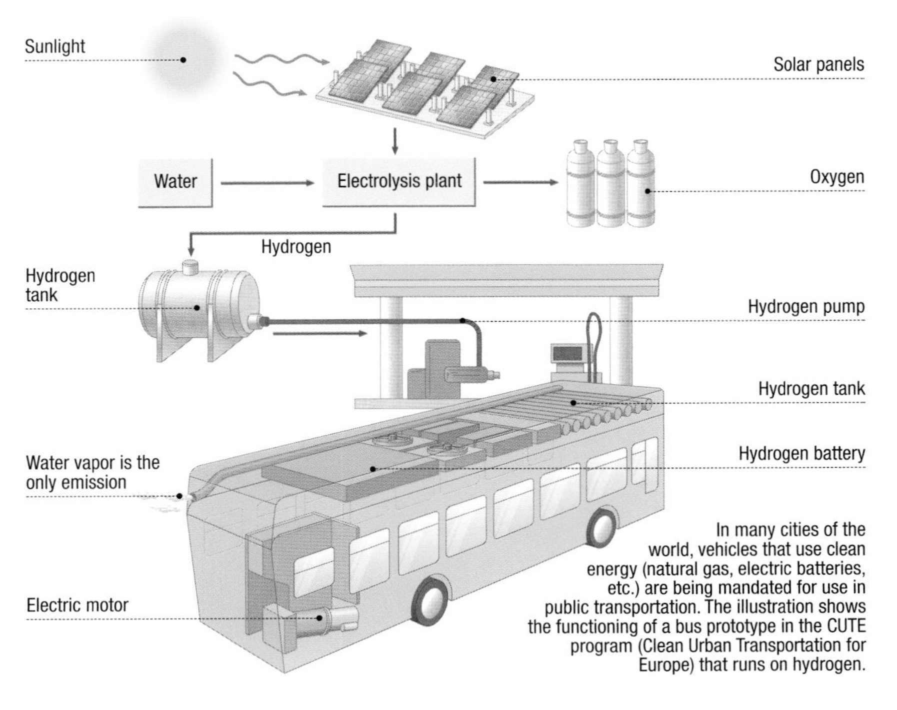

In many cities of the world, vehicles that use clean energy (natural gas, electric batteries, etc.) are being mandated for use in public transportation. The illustration shows the functioning of a bus prototype in the CUTE program (Clean Urban Transportation for Europe) that runs on hydrogen.

Public transportation (subway, bus, streetcar, etc.) not only takes up less room and generally produces less pollution, but it is also much cheaper for the user.

AIRPLANES AND SHIPS

For travel between countries or cities more than 185 miles (300 km) apart, air transportation is commonly used because of its speed. Throughout history, ships have been one of the main methods of transportation between costal areas, although nowadays they are used increasingly for transporting goods instead of passengers. Both airplanes and ships pose problems for the environment. The former produce tremendous noise and burn lots of fuel. The latter often dump residue and fuel into the ocean, and that causes pollution of marine environments.

One drawback to airplanes is that they require a colossal infrastructure that has an irreversible effect on the landscape.

MAIN WORLD AIRWAYS

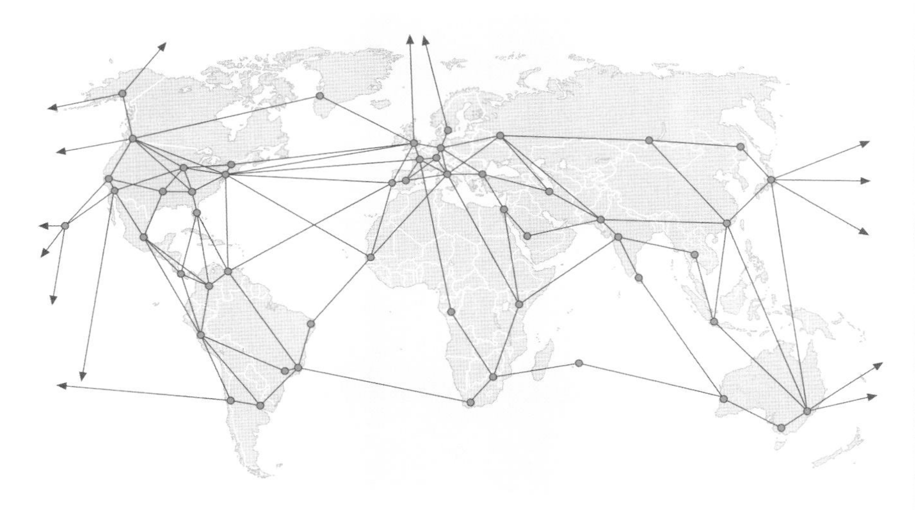

Although it hardly seems possible, air space is saturated with planes traveling back and forth every day.

A DAY WITH NO AUTOMOBILES

Throughout the year there are days when people do things to recognize certain groups or call attention to special problems. One of these focuses on the problem of too many automobiles running through certain areas of large cities. This is complemented by celebrations that include walking or riding bicycles, demonstrations, and other activities.

PEDESTRIAN ZONES

This involves a series of streets located in commercial areas. These streets are closed to traffic all year, except for special vehicles (e.g., ambulances, police, street cleaning, and pickups and deliveries).

FRESHWATER POLLUTION

Ever since prehistoric times, human populations have clustered around rivers and freshwater lakes, because water is one essential component for life. The development of agriculture and livestock industries also requires the presence of water nearby. But water use does not stop there, for sadly, rivers have been used, and still are used, as sewers for dumping waste waters and industrial residues.

HOW RIVERS BECOME POLLUTED

River pollution occurs mainly through dumping toxic or nondegradable substances. Sometimes we learn through the newspapers or television that there has been a sudden die-off involving a great number of fish in a river. In this instance, it is most likely that there has been some **dumping** of **toxins** from a factory. Another type of very common pollution in rivers all around the world is the **purines** and other residue from livestock, which cause eutrophication of the water.

Industrial dumping (from paper, textile, and chemical plants, for example) is one of the main sources of river pollution.

PURINES

This is pig manure. It is an extremely harmful residue for the environment, and it is very difficult to eliminate, because treatment involves high economic and energy costs.

The best way to avoid pollution is not to dump substances that are harmful to rivers and to make industries treat their residues in special plants, and requiring that urban wastewaters pass through purifiers before reaching a river.

Farm wastes have to go through a recycling plant to prevent soil contamination.

EUTROPHICATION

Eutrophication is the main problem with freshwater ecosystems. Eutrophication of the water results when too much phosphorus and too much nitrogen are added. Then the populations of microscopic algae, which use these substances as a basic food, experience explosive growth and color the water. This color is an opaque barrier that keeps sunlight from reaching the depths of the water and that affects the growth of underwater weeds at that level. Because the weeds receive no sunlight, they cannot generate oxygen through photosynthesis, and, as a result, the environment lacks this substance, which is vital to animal life.

The phosphorus and nitrogen in a river come from excess fertilization of crop fields and an excessive concentration of livestock; this results in a large quantity of artificial nutrients.

DIAGRAM OF THE EFFECT OF EUTROPHICATION

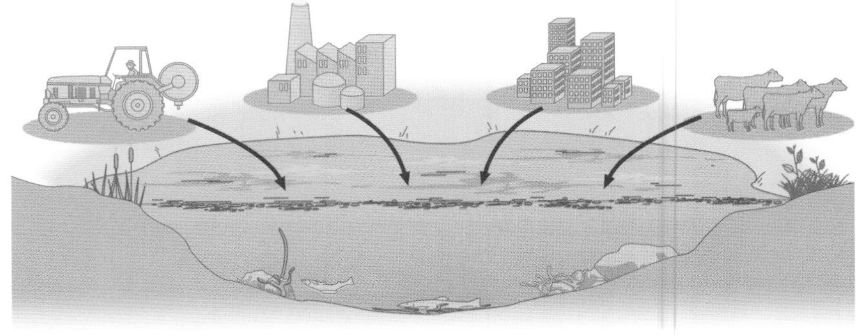

A QUESTION OF VOLUME

Freshwater ecosystems are extremely fragile in comparison with land and marine ecosystems. That is basically because:

1. Most substances dissolve in water, so they enter more easily into the bodies of aquatic organisms.

2. The volume of fresh water in any river or lake is much lower than that of a sea or ocean, and that makes it easy to alter the nature of the water that the fish and invertebrates so depend on.

Volume of contaminants

Volume of river water

Volume of contaminants

Volume of ocean water

A spill in a river has greater consequences than in the sea, because the seriousness of the toxicity depends on the concentration of the toxin in the water. In fact, the proportions are greater than indicated in the illustration, but the volume of the ocean could not be represented on the page.

Lakes high in the mountains are extremely poor in nutrients, and that's why they are so crystal clear. The organisms that live in them are very sensitive to any change. As a result, these lakes must not be used for bathing.

Another type of harm done to rivers is the regulation of current by means of dams. These huge engineering projects impose a physical barrier to fish and other creatures whose ecology depends on migration between the headwaters and the mouth of the river.

THE EFFECT OF SALINIZATION OF AN AQUIFER

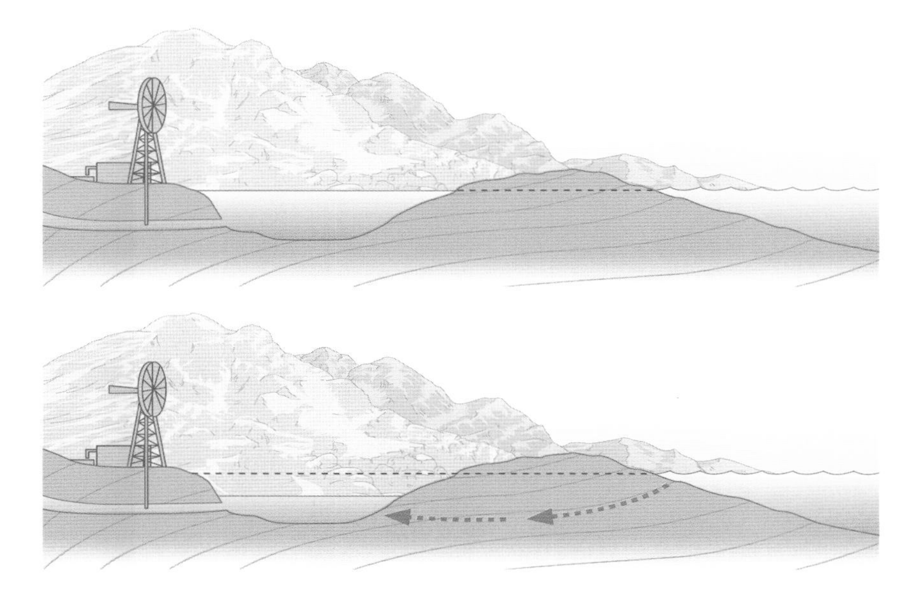

SALINIZATION

A serious problem is developing along the seashore. Aquifers are being overexploited in such a way that seawater is getting into them and making the water salty.

PROBLEMS WITH SALTY WATER

For health	Not suitable for drinking.
For agriculture	It burns plants.
For industry	It rusts machinery.
For the ecosystem	It destroys vegetation and kills animals.

THE POLLUTION OF SEAS AND OCEANS

More than 70% of the Earth's surface is covered by water, and most of that is **ocean water**. As a result, our seas and oceans are an important, **vital** resource, and we should take care of their health. Unfortunately, marine pollution affects the whole world today, and, although the oceans are a huge environment that can absorb a great quantity of pollutants, a day will come when their capacity for **self-purification** becomes exhausted and then there will be no turning back.

THE LAST LINK IN POLLUTION

Sooner or later, all the materials that are discarded or spilled into **land ecosystems** end up in the ocean by means of the rivers. If you look at the seashore after a storm, you probably will note the brown color of the water. This is caused by the great quantity of earth and **sediments** that have been carried into it. Also, many times the water is filled with plastic, pieces of wood, and all kinds of floating materials that end up accumulating on the beaches or the rocks along the coast. Thus, to keep the sea from becoming polluted, we must begin by ceasing to contaminate the land surfaces and the rivers.

Dolphins, tortoises, and other animals swallow pieces of plastic that float in the water and that causes them to suffocate or to suffer from intestinal blockage; in either case they experience a painful death.

"THE SEA RETURNS EVERYTHING THAT DOES NOT BELONG TO IT."

This famous saying by people who live along the coast is not entirely accurate, because the sea bottom along the shore is often a dump for all kinds of objects; we just cannot see them from the surface.

Every year there are campaigns to clean up the beaches. The government takes care of this in tourist areas; however, in areas with no commercial interest but with natural beauty, many times the labor falls to ecology groups. In these operations tons of trash are picked up that have been deposited by the waves after traveling various distances on the ocean currents.

INVISIBLE POLLUTION

In addition to **household** and **agricultural pollution** that is carried by the rivers, there are also **heavy metals** and other toxic by-products that are dumped directly into the ocean by industries located along the coast, as well as refuse from many coastal cities that dump their wastewaters directly into the ocean without purifying them. This pollution is diluted in the water and seems to disappear; however, many analyses show that in fact the pollutants are accumulating slowly but surely in the life forms that inhabit the oceans.

Every year many cetaceans become beached along the coasts and end up dying. The causes are not well known, but there are indications that marine pollution (chemical, acoustic, etc.) is the cause of these disasters. It is presumed that these agents disturb the echolocation sense of these animals; when they become unable to orient themselves, they end up running aground on the coast.

POLLUTION DEPOSITS

Coastal mollusks (mussels, clams, etc.) are organisms that act as filters; thus, they incorporate heavy metals, such as mercury, cadmium, lead, and chrome, into their bodies. In that fashion they introduce these poisons into the marine food chains.

The ice of the Arctic and Antarctic has been analyzed, and it has been shown that pollutants are accumulating even there.

TANKER TRAFFIC

This is one of the most serious threats to **marine ecosystems**. This involves huge ships that carry thousands of tons of crude oil. When one of them runs aground or sinks (because many of these ships are in very poor condition), the result is an ecological disaster with devastating consequences. The **crude oil**, which floats on top of the water because it is less dense, reaches the shore and covers the beaches and rocks, turning them a depressing black color. Organisms are poisoned and die, or they die of cold because the oil destroys their layer of insulation; the seaweed is robbed of light and dies, and so on. In addition, petroleum takes many years to degrade.

THE DISASTER OF THE *PRESTIGE*

One of the most recent **black tides** was the one in 2002 and 2003 that affected the northern coasts of Spain and Portugal and the southeastern coast of France. It was produced by the sinking of the monohull tanker *Prestige.* Cases like this one are the result of negligence on the part of petroleum transport companies that fail to invest in repairs and maintenance of the ships they use, in an attempt to save money.

When a black tide occurs, not only the natural environment is affected, but the coastal economy based on fishing and tourism also is totally destroyed.

THERMAL POLLUTION

The rise in the temperature of the ocean everywhere on the planet is evidently the cause of death for the coral in many atolls, islands, and barrier reefs.

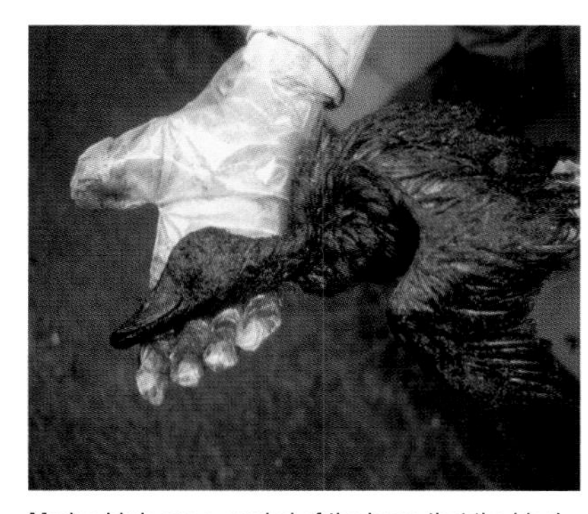

Marine birds are a symbol of the harm that the black tides produce. Covered with petroleum, they cannot regulate their body temperature, and they die of cold. The petroleum also enters their respiratory and digestive tracks, and they die from poisoning.

SMALL SPILLS

Many times motorboats spill small amounts of fuel; when these amounts build up, they too cause serious damage to marine ecosystems. In addition to this, many ships clean their tanks on the high seas.

TOURISM

Although tourism is not a type of **pollution** per se, it has been one of the major causes of the **degradation** of coastal **marine ecosystems** throughout the world. The construction of hotels and developments along the coast and long seaside avenues has caused the disappearance of many species in places where they used to nest or search for food. In addition to the physical disappearance of terrain appropriate for fauna and flora, tourism also produces greater amounts of urban wastes, motorboats leak oils and other pollutants into the sea, and causes lots of noise pollution, and so forth.

Just like industrial plants, large tourist concentrations on the coast lead to dumping of pollution; that is why wastewater treatment plants are so important.

UNDERWATER ACTIVITIES

Scuba diving also can damage the marine environment if it is not practiced properly. For example, people hit the **coral** and other fragile bottom dwellers with their flippers and interfere with recolonization, and the air bubbles given off from the oxygen tanks alter the ecological conditions in **underwater caves**, and more.

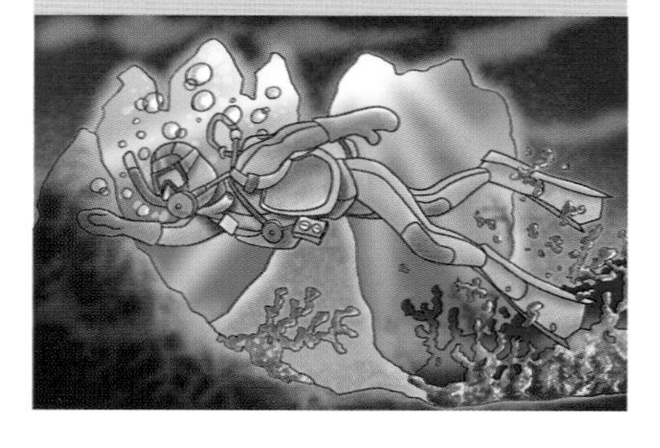

Underwater spear fishing has caused many species, such as the grouper, to disappear or become rare in many places.

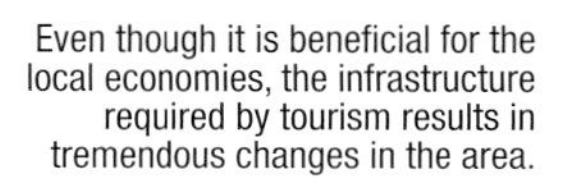

Even though it is beneficial for the local economies, the infrastructure required by tourism results in tremendous changes in the area.

NUCLEAR TESTS

Some countries carry out **nuclear tests** on coral islands that belong to them for historical reasons but are located far away from metropolitan areas. As we would expect, these tests are very harmful to the **flora** and **fauna** that live there; they end up suffering all the symptoms of nuclear contamination (death, illness, genetic alterations that last for many generations, and so forth).

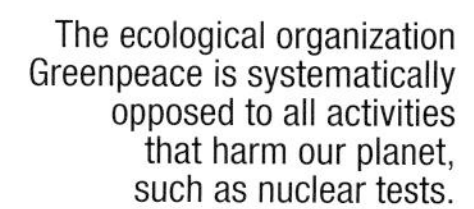

The ecological organization Greenpeace is systematically opposed to all activities that harm our planet, such as nuclear tests.

Nuclear power plants located on the coasts use seawater for cooling the reactors. The water they return to the ocean is much warmer than before and that causes profound changes in the coastal ecosystems.

The proliferation of motorboats and their improper use and uncleanliness cause much harm to coastal waters.

A TIME BOMB

Containers of **nuclear waste** dumped into the sea already have cracks in them, and some have broken open, releasing dangerous pollutants.

THE IMPORTANCE OF CURRENTS

Marine currents are very important to the degree of pollution in the ocean. To illustrate, we can compare the **Mediterranean Sea** to the **Baltic**. In the first instance the connecting point with the **Atlantic** Ocean is the Straits of Gibraltar. At this point the surface currents go from the ocean toward the Mediterranean, while the deep currents flow in the opposite direction. This is important because the deep waters are the ones rich in nutrients (and pollutants), so the Mediterranean gets rid of the ones it has too much of and receives fresh water; as a result, its waters are relatively transparent and clear. The opposite occurs with the Baltic: nutrients and pollutants accumulate in it, and its waters are more green in color.

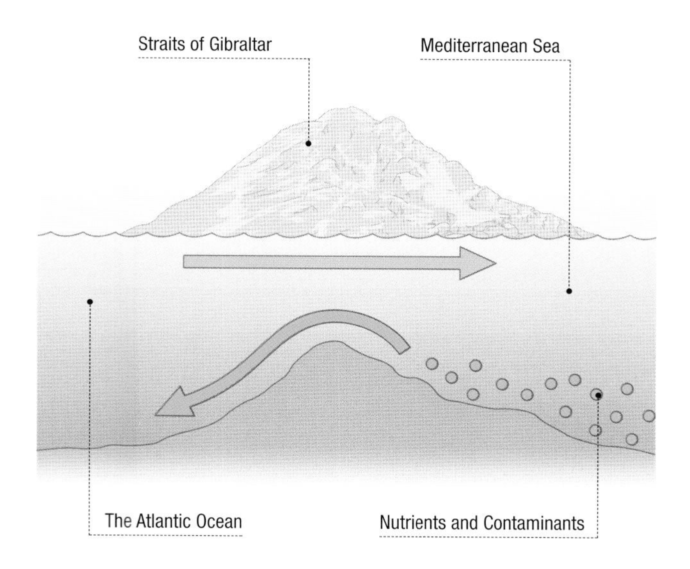

Urban wastewaters contain a large quantity of bacteria that can produce gastric and dermatological problems and eye and ear infections if they are discharged into the ocean near a beach frequented by swimmers.

THE MAIN DISASTERS CAUSED BY OIL TANKERS

Year	Tanker	Location of the disaster
1967	*Torrey Canyon*	Off the south coast of Ireland
1975	*Jakob Maersk*	Off the coast of Portugal
1976	*Urquiola*	La Coruña (Spain)
1978	*Amoco Cadiz*	English Channel
1979	*Atlantic Empress*	Caribbean Sea
1979	*Ixtoc One*	Gulf of Mexico
1989	*Exxon Valdez*	Alaska
1992	*Aegean Sea*	La Coruña (Spain)
1999	*Erika*	Off the coast of Brittany (France)
2002	*Prestige*	Off the coast of Galicia (Spain)

PURIFYING AND MANAGING WATER

In contrast to the other **animals** on the planet, **humans** are not content to live with what they can get from **nature**. At home they wash clothes very frequently, people take showers every day, wash their dishes, and so forth. This life style creates a great need for water, plus such a great amount of waste that nature is incapable of taking it into its **natural cycles** and integrating it. Agricultural and industrial wastes are added to these domestic uses and waste production.

THE WATER CYCLE IN SOCIETY

In many countries in which precipitation is not reliable, there are many dams that store up the waters of the rivers. This water is used for agriculture (and for producing electricity), but it is not of the quality necessary to be considered **potable**, so it has to be treated in **purification plants**. From there it is distributed to houses and industries; when they use it, they dirty it, and fill it with wastes. This **wastewater** is collected by means of pipes—the sewer system in this case—and is brought to **water purification plants**, where many of these wastes are removed. The purified water is then returned to the environment (river or sea).

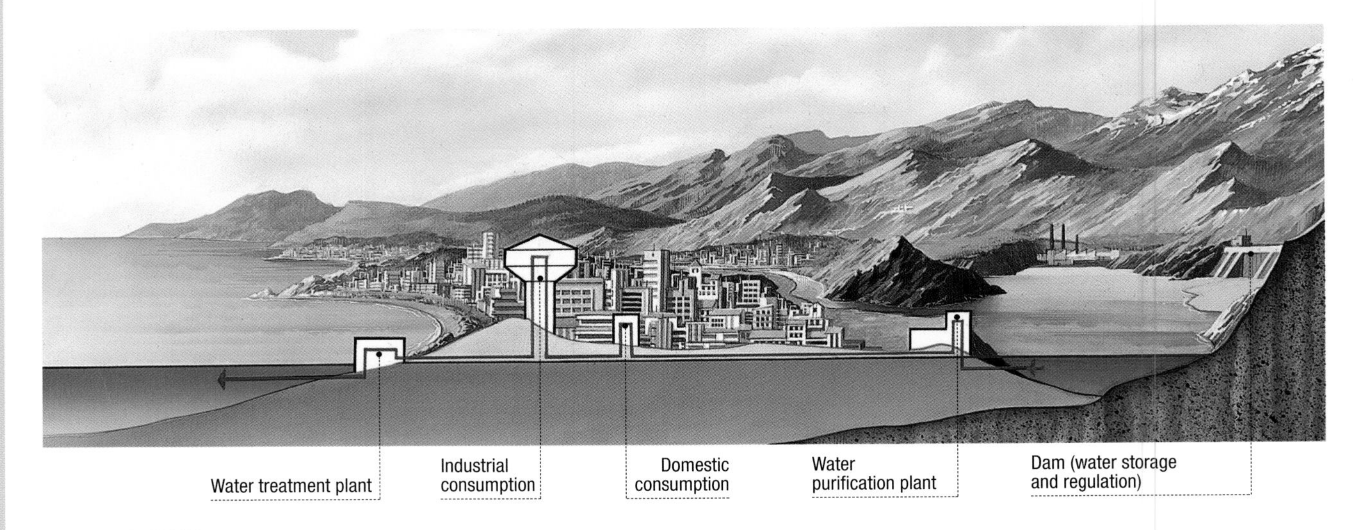

PRODUCING POTABLE WATER

Potable water does not come only from water purification plants. In many places there are still private **wells** that are kept in good condition. However, they are decreasing in number because of ground pollution and excess exploitation of water resources in the subsoil.

In areas located near the ocean, desalinating plants are used to convert seawater into fresh water.

Excessive use of wells causes them to go dry or become salty.

Only 3% of the water on the planet is fresh and usable by humans.

TYPES OF WATER PURIFIERS

There are several models of **water purifying plants** that are used for treating one type of water or another. Most of the wastes from **households** are not too toxic for the environment, and in fact, nature is able to eliminate them on its own. The problem is that they are generated in such tremendous quantities that it is difficult for the environment to absorb it all. In **industry**, on the other hand, numerous toxic substances are generated. For them, a water purification station condenses the process that would take place in nature. There are two basic types of purifiers: **biological** and **chemical**.

BIOLOGICAL PURIFIER

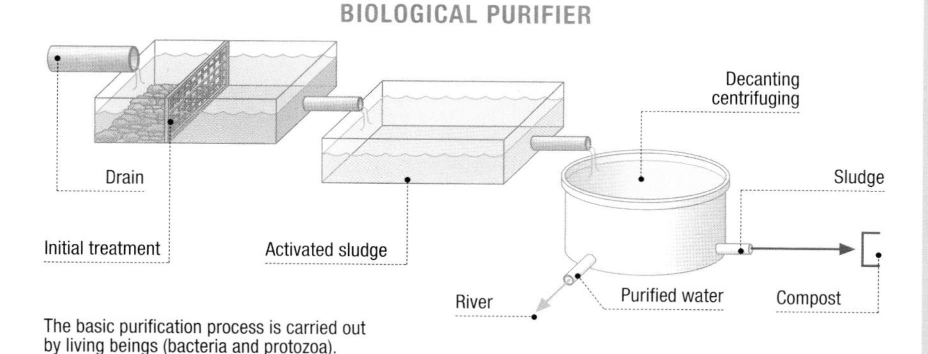

The basic purification process is carried out by living beings (bacteria and protozoa).

ACTIVATED SLUDGE

This involves millions of microorganisms (bacteria and protozoa) that eliminate all the organic matter from the wastewater. They feed on it and convert it to the organic matter of their own bodies.

CHEMICAL PURIFIER

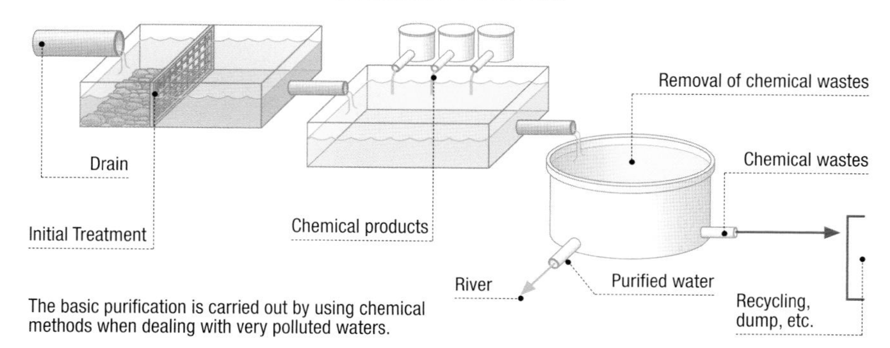

The basic purification is carried out by using chemical methods when dealing with very polluted waters.

We need to take care of freshwater environments not only for ecological reasons, but also for our own survival.

The rational use of water requires some sacrifices, but the main thing is not to waste it.

WATER SCARCITY

Freshwater is a scarce commodity that we must take good care of if we do not want to run out of it some day. For centuries, in areas where water was once plentiful, people did not give it the importance it really deserves, and they overexploited the **aquifers** and polluted the rivers without considering that one day there might not be any more water. That is the state we have arrived at today.

WATER USAGE

Appropriate	Inappropriate
Keep faucets turned off.	Leave the water running.
Repair the plumbing.	Ignore leaks in the plumbing.
Water sparingly.	Use a hose for watering.
Grow native plants in regions where water is scarce.	Grow plants that require moisture in areas where water is scarce.
Recycle water.	Use water only once.

AIR POLLUTION

The composition of the **atmosphere** that surrounds the Earth is not at all constant; instead, it depends greatly on the relationships it establishes with the oceans, with the mineral components of the planet, and, especially, with life forms. Until scarcely two centuries ago, these relationships were balanced so that there were no great changes in the composition of the **air**; however, modern life, and especially industrialization, has changed the global panorama, and the composition of today's atmosphere is changing because of the **emission** of many gases.

A FLUID MEDIUM

Just as with seas and oceans, the **pollution** that reaches the atmosphere is diluted by its great volume. This is the main difficulty in raising society's consciousness so that it limits harmful **gaseous emissions** into the environment, because in the short run the effect of air pollution is scarcely noticeable. However, day after day, the pollutants are becoming concentrated to the point that the adverse effects are already being seen in some places, and, on a worldwide scale, it is becoming clear that **global warming** is beginning to affect the overall climate.

MAIN ATMOSPHERIC POLLUTANTS

Pollutant	Causative agent		
	High	**Medium**	**Low**
Sulfur oxides	Industry	Homes	Vehicles
Nitrogen oxides	Vehicles	Industry	Homes
Particles in suspension	Vehicles	Industry	Homes
Carbon dioxide	Vehicles	Industry	Homes
Lead	Vehicles	Vehicles	Industry

A look at the outside of buildings easily confirms air pollution in cities.

Gaseous pollutants react with elements in the atmosphere to produce compounds even more harmful than the original ones.

ACID RAIN

Nitrogen and **sulfur** oxides may react with water vapor in the atmosphere and create **nitric** or **sulfuric** acid. When these acidic compounds (mixed with fog, snow, or rainwater) fall onto plants, they burn them; when they fall onto monuments they erode them; and when they fall into freshwater, they make it acidic, thereby causing the death of aquatic life forms and also harming human life.

THE ACID RAIN PROCESS

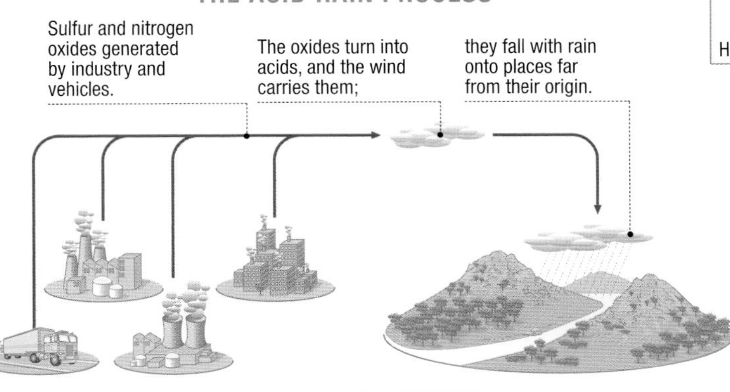

THE PROCESS OF AIR POLLUTION

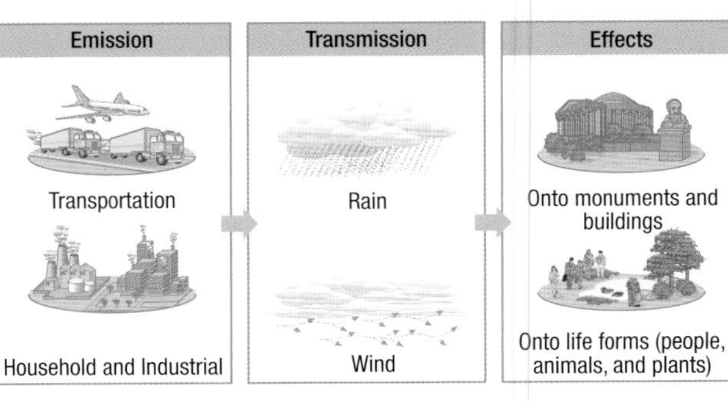

AUTOMOBILE CATALYTIC CONVERTERS

Catalytic converters convert carbon monoxide, nitrogen oxides, and hydrocarbons, all of which are very toxic, into nitrogen, carbon dioxide, and water vapor, which are all harmless to health.

THE FUNCTIONING OF A CATALYTIC CONVERTER

Gaseous pollutants: nitrogen oxide, Carbon monoxide, Hydrocarbons

Purified gases: Nitrogen, Carbon dioxide, Water vapor

Motor

Exhaust

Precious metals in catalytic converter

THE PURIFYING EFFECT OF A TRIPLE CONVERTER

(emissions in g/km for an automobile traveling at 35 miles/50 km per hour)

with catalytic converte

without catalytic conv

Carbon monoxide

Hydro-carbons

Nitrogen oxide

CARBON DIOXIDE (CO_2)

Carbon dioxide alone is not a pollutant, because it is a very common gas in the atmosphere, and its **cycle** is very closely linked to **life forms**: plants absorb it and incorporate the carbon into their bodies by means of **photosynthesis**, and all life forms release it through **respiration**. Still, today's carbon dioxide emissions are so great that plants cannot absorb them all and it all accumulates in the atmosphere, which causes **global warming** on the planet.

CO_2 reaches the atmosphere mainly through the respiration of life forms and through the combustion of organic matter (wood, plants, hydrocarbons, and so forth).

Forest destruction makes it even more difficult to balance the amount of carbon dioxide in the atmosphere.

THE PHENOMENON OF TEMPERATURE INVERSION

Cold air
Warm air
Thermal barrier
SO_2
NO
Pb

In Australia and Indonesia, during the summer, there are often huge forest fires that send up dense clouds of smoke. The smoke stagnates in adjoining areas and forces people to wear face masks to breathe.

POLLUTION IN CITIES

In some cities located on low terrain, a meteorological phenomenon known as **temperature inversion** occurs in the winter. A stable mass of cold air settles on a city and forms a type of thermal barrier that prevents the escape of the gases generated by traffic, heating, and factories. If this situation lasts for several days, the atmosphere in the city becomes too loaded with pollutants, and people's health may be endangered. This situation is resolved when new winds and rains arrive. . . or when the sources of **pollution** are eliminated.

OZONE

This is a very valuable compound in the upper layers of the atmosphere, but it is very dangerous in the lower layers, since it produces cancer and other serious transformations in life forms.

Many times the air in large cities becomes unbreathable.

EMISSIONS AND FOREST REGENERATION

Human societies have been altering and abusing the **environment** for centuries, but an interest in protecting it arose at the start of the 1970s. After arriving at an excessive state of **ecosystem degradation**, the industrialized nations that cause the majority of polluting emissions decided to work together on this task. This collaboration is not the fruit of altruism or a sense of guilt, but instead, an essential need to safeguard the **quality of life** for present and future generations.

INTERNATIONAL COOPERATION

Most of the **industrialized nations** have conducted numerous campaigns and held meetings to establish time limits for reducing **emissions of pollutants**, but so far few of them have really met these commitments. The main problem is that it is expensive to care for **ecosystems** (although it is much more costly to repair them), and it requires a profound change in firmly ingrained work practices.

THE PRINCIPAL INTERNATIONAL PROCEEDINGS

1976	A coordinating commission for information on the ozone layer is established by the United Nations
1985	A summit is held in Vienna in which the first international declaration of good intentions for protecting the ozone layer is signed.
1987	*The Montreal Protocol*: a commitment to reduce emissions of chlorofluorocarbons, which destroy the ozone layer, by 50% by the year 2000
1992	Rio Summit or Conference
2002	Johannesburg Conference

The most appropriate alternative for reducing harmful emissions is the use of **renewable energy** (solar and wind). However, these are still in the development stage.

Eolian power (the word comes from the Greek Eolo, the wind god) is clean and inexhaustible, even though it causes some visual pollution and the units can harm birds.

Present reserves of natural gas are adequate to meet the needs of the next hundred years. The photo shows natural gas storage tanks.

A Pier for methane tankers (Tarragona, Spain).

NATURAL GAS: A LESS POLLUTING FUEL

Natural gas is now the most common fuel for **heating** and producing heat in countless industrial processes. Compared with other **fossil fuels**, it generates a much lower level of waste per unit of energy. For example, the amount of CO_2 emitted through natural gas combustion is half the amount as with carbon and petroleum; emissions of nitrogen oxides is between a third and a half, and sulfur oxides are less than a hundredth. It also does not produce any particles in suspension. The expanding use of this energy source has made a significant contribution to reducing **harmful emissions**.

Natural gas is composed primarily of **methane**. It is a by-product of the decomposition of organic matter located between rock layers. Once it is extracted from the deposit, it is conducted through pipelines to distribution plants, which supply it to homes and industries.

DEFORESTATION AND THE ATMOSPHERE

The problem of harmful gas **emissions** in the atmosphere is aggravated by the destruction of the tremendous areas of **tropical rainforest** located in the latitudes near the equator, which are known as the lungs of the world. Thousands and thousands of acres of forest disappear every day for mainly economic reasons, for they are exploited for their **wood**. With each tree that is lost, the capacity for absorbing the excess **carbon dioxide** that accumulates in the atmosphere further decreases. It is nearly impossible to regenerate these millenary forests, because they grow on poor soil where seeds cannot germinate.

THE EFFECTS OF DEFORESTATION

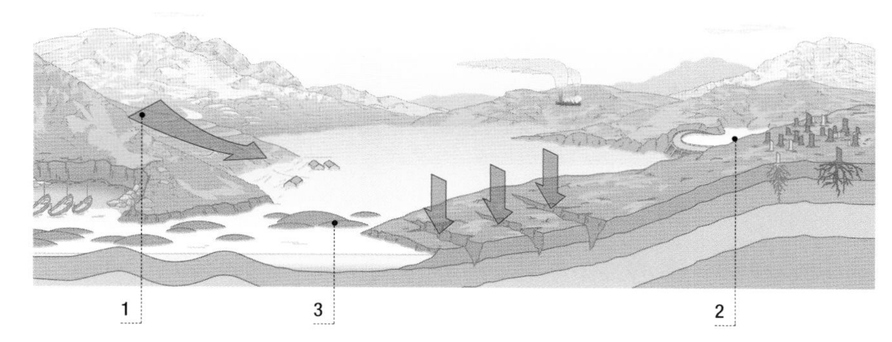

(1) Rainwater quickly runs down the hillsides, leaving the soil and settlements without protection (resulting in damaged crops, desertification, etc.); (2) The earth swept away fills up the reservoirs, reducing their capacity, and the river, causing floods; (3) the sediments carried downstream create new islands, change the underwater bottoms, and reduce fishing areas.

Fires are another major cause of forest loss. The most serious ones occur in places with a dry climate where forests still exist, such as Australia.

THE MAIN CAUSES OF DEFORESTATION IN THE TROPICAL RAINFOREST

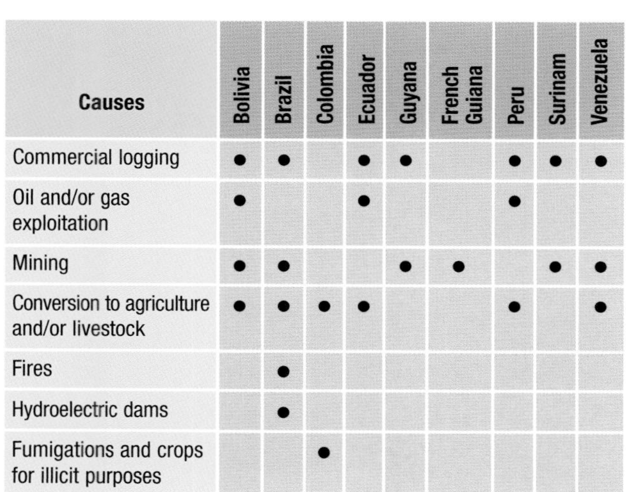

Causes	Bolivia	Brazil	Colombia	Ecuador	Guyana	French Guiana	Peru	Surinam	Venezuela
Commercial logging	●	●		●	●		●	●	●
Oil and/or gas exploitation	●			●			●		
Mining	●	●			●	●		●	●
Conversion to agriculture and/or livestock	●	●	●	●			●		●
Fires		●							
Hydroelectric dams		●							
Fumigations and crops for illicit purposes			●						

BE CAREFUL WITH FIRE

1. Keep fires under control and burn no trash.
2. Never toss out lit cigarettes.
3. Do not use fireworks in risky areas.
4. Do not burn plant wastes.
5. Leave no trash or debris in the forest.
6. Cook only in a proper barbecue.

Today in industrialized countries, forests are considered a very valuable commodity for society. As a result there are intensive campaigns for forest regeneration.

THE EVOLUTION OF THE EUROPEAN FORESTS

Formerly, the entire European continent was nearly totally covered by **forest**. Starting in the Middle Ages there was massive **cutting** to create lands and use them for agriculture and livestock, and to get firewood and wood for construction. This caused the near total destruction of the original forested layer.

Nowadays the process has moved in the other direction. The abandonment of crop fields is increasing the treed area, which is occupying the original terrain. This is the **natural regeneration** of the forest. The process is also being helped along through **forest replanting**.

REGENERATION OF BURNED AREAS

To regenerate a burned forest, we begin by planting seeds or small seedlings of native species that can grow in open spaces. That way natural generation is encouraged.

OTHER TYPES OF POLLUTION

When we speak of **pollution** it is most common to think of harmful gas emissions in the atmosphere, spills in rivers and seas, solid wastes, and so forth, in other words, of pollution caused by materials and substances. However, there are other types of pollution that could be classified as physical; these do not involve a spill into the environment, but instead a physical effect (a warming, a noise, etc.) that alters the proper functioning of the **ecosystems**.

NUCLEAR POLLUTION

Half a century ago people discovered how to release a tremendous amount of energy from the destruction of the **atomic nucleus** of certain elements (uranium). This energy at first appeared to be clean, because, unlike fossil fuels, it released no **emissions** into the atmosphere. However, the wastes produced are extremely dangerous to the health of all life forms, because the radiation given off can burn and alter **cells** and **genes**. In addition, they remain active for thousands and thousands of years. They have to be kept in hermetically sealed containers to keep them from getting out, but it is nearly impossible to find a container that will last as long as most of the nuclear wastes will remain active.

Although nuclear power plants theoretically provide clean energy, their residues remain intensely radioactive for a long time and create an enormous problem regarding their disposal.

Nuclear power plants use the heat created in atomic fission to generate electricity.

NOISE POLLUTION

Noise is one very special type of pollution. It is a physical agent, not a material that is introduced into the environment, and it can be considered pollution only when it exists in excessive quantity. Then it can affect the nervous system of animals and cause stress. The effects of excess noise are cumulative; for example, if you listen to a loud noise one day there are no consequences, but if you listen to it for a prolonged period of time it can cause **deafness**.

Motors generally are the main cause of excess noise in natural environments. These same noises can also affect the health of city residents.

The noise of cities is often unbearable, so there are frequent campaigns to reduce it.

NUCLEAR FUSION

There is a new technology that consists of producing energy from the fusion of some of the hydrogen isotopes in water. This produces lots of energy, and the waste products are harmless, but this is still in the research stage.

When a vehicle goes through the forest the animals run away. In ecosystems where there is too much noise many species have disappeared.

HEAT POLLUTION

Many processes in some industries, and especially the **nuclear** and **thermal power plants**, generate a great deal of heat. A large part of this heat is released into the environment, creating special conditions around the emission point. In general, the area affected by the temperature increase is small. Even though the heat dissipates very quickly, its consequences are evident: a gradual disappearance of the organisms that live there, followed by colonization by others from a warmer climate.

Thermal power plants use coal or oil and produce significant pollution if the appropriate filters are not used.

Large cities experience heat pollution because of the use of heating systems and through the accumulation of industrial and household gases that absorb heat and hold it in the surrounding atmosphere.

Certain studies indicate that excessive use of cell phones can cause harm to the user over the medium and long term.

ELECTRICAL POLLUTION

There is a big debate in society over the effect that **high-tension lines** have on living organisms. Some people assert that there is a higher incidence of cancer in populations that live near these electrical lines; however, these conclusions have not been proven scientifically.

TELEPHONE ANTENNAS

The problem of **electromagnetic fields** generated by the large **relay antennas** for cell phones is very similar to the one involving **electric fields**. So far the studies done on that subject conclude that they are not harmful to health. However, people generally prefer that they not be installed near their homes.

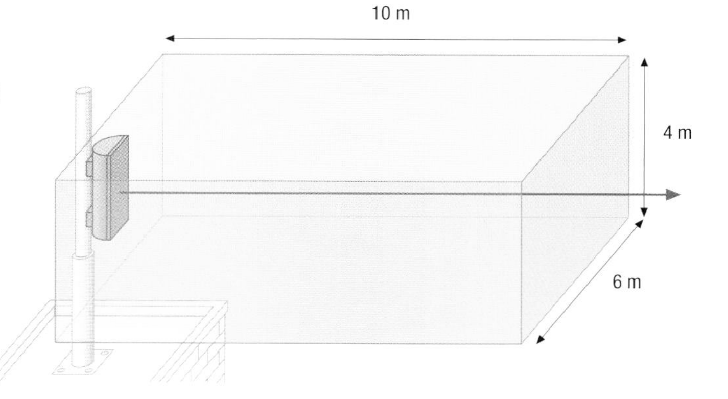

Safety zone that must be observed near a cell phone antenna between 100 and 1,000 watts of broadcast power.

THE GREENHOUSE EFFECT AND OZONE

The **atmosphere** is a fluid layer that surrounds the Earth and whose composition and characteristics are indispensable for life on the planet. In contrast to what many people think, it is a living layer in which there are continuous reactions and interactions with life forms but that in the end remains in balance. Still, during recent centuries **human activities** have upset this equilibrium by producing some real problems that one day may cause the extinction of our species.

WHAT IS THE GREENHOUSE EFFECT?

The **Sun's** rays that reach the surface of the Earth bounce off the ground and are returned to space in the form of **infrared radiation**. Some molecules in the atmosphere absorb this radiation and warm up. The atmosphere thus acts like a blanket that prevents **heat** loss. This is what is known as the **greenhouse effect**, which is a natural phenomenon without which there would be no life, since the temperatures would be too low for life forms. The problem arises when humans, with their carbon dioxide (CO_2) emissions and other compounds (methane, water vapor, nitrogen oxides, and so on), alter the balance and cause an increase in the greenhouse effect and an exaggerated temperature rise in the atmosphere.

THE GREENHOUSE EFFECT

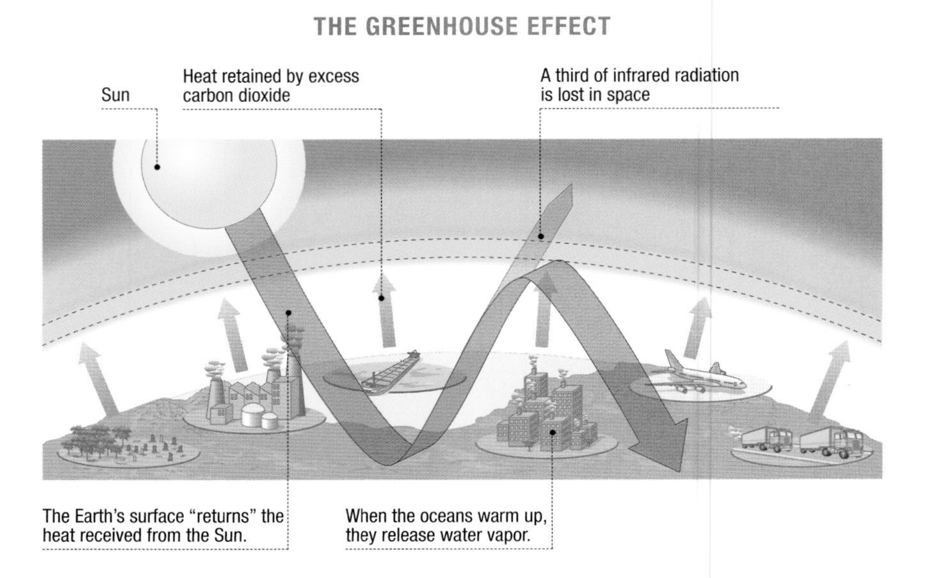

AN INCREASE IN THE EARTH'S TEMPERATURE

Excess emissions from factories, automobiles, and other sources is one of the major factors that contribute to the greenhouse effect and the warming of the atmosphere all over the planet.

CO_2 is produced through the respiration of life forms and the combustion of organic products such as wood, coal, and petroleum and its derivatives.

This is a mean value for the entire planet, because in some places the temperature has gone up several degrees and in others it has gone down. This causes large-scale **climatic changes**, which produce serious natural disasters.

The problem of the increased greenhouse effect is that life forms cannot adapt to the new climatic conditions as quickly as they occur, so species disappear at a much faster pace than normal.

PROJECTION OF THE TEMPERATURE INCREASE OF OUR PLANET

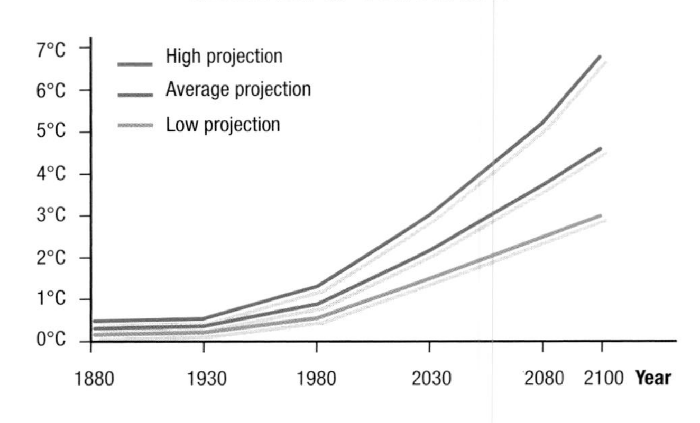

THE HOLE IN THE OZONE LAYER

In the 1980s people began speaking of the **hole in the ozone layer** over Antarctica. This hole was a clear symptom of the planet's deterioration because of excess **atmospheric pollution** and a serious danger for the survival of the life forms that inhabit it. Then there came a new movement to protect the environment, and international conferences were organized to discuss the topics that affect the planet in general and to require corrective measures from all countries.

THE MONTREAL PROTOCOL (1987)

This meeting was the first to deal seriously with the problem of the disappearing ozone layer. It is not decreasing evenly all over the planet, but instead in the form of a large hole over the poles, which increases dramatically during the springtime.

The central spot located above Antarctica represents an area where there is no ozone and that thus allows the entry of radiation dangerous to living beings. This is one of the few environmental subjects that the nonscientific press has covered extensively, which is a sign of its importance.

CHLOROFLUOROCARBONS (CFC)

These are the main enemies of ozone molecules. They are compounds that contain chlorine atoms; when they reach the ozone layer they react with it and destroy it. A single molecule of chlorine can destroy hundreds of ozone molecules.

UTILIZATION OF CFCs

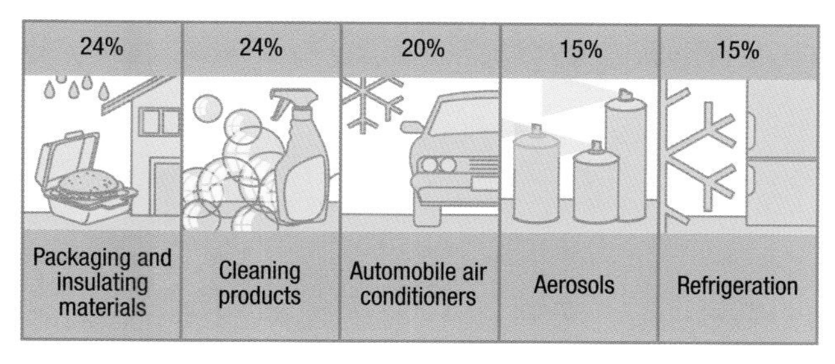

24%	24%	20%	15%	15%
Packaging and insulating materials	Cleaning products	Automobile air conditioners	Aerosols	Refrigeration

OZONE

This is a molecule made up of three **oxygen** atoms. In the atmosphere, at an altitude of about 82,000 feet (25,000 m), there is an especially rich ozone layer whose main characteristic is that it absorbs **ultraviolet rays** and keeps them from directly reaching the surface of the Earth. This is very important to living beings, because this radiation causes changes in cells, producing diseases such as **cancer** and serious **mutations** in the genetic material of animals (including humans) and plants, thereby often interfering with their reproduction and reducing their natural defenses.

It is estimated that a loss of 10% of the ozone layer over the Earth would cause a 30% increase in skin diseases.

CFCs practically do not exist in nature. They are a man-made product used in aerosols, refrigeration liquids, microcircuit cleaning products, foam packing materials for fast food, and so forth.

In areas where the ozone layer has experienced significant degradation, such as Australia, bathers are prone to experience serious harm if they are exposed to too much sun without protection.

ALTERNATIVE ENERGY SOURCES

Some estimates conclude that natural **resources** will not last forever. However, it has been calculated that they will last long enough for the **emissions** produced in burning them up to exceed what the Earth's ecosystem can withstand. Thus, it is crucial to begin using the new **energy sources** that are referred to as clean, because they do not emit gases or other byproducts that are toxic to the environment. In fact, alternative energies are limited to transforming an energy source from nature (sunlight, wind, etc.) into another form of energy that humans can use.

ALTERNATIVE ENERGY SOURCES

These are the only option for **sustained development** so that in the future there will be **ecosystems** like the ones we have in our time. However, systems of destructive development continue to prevail on our planet: the worldwide fleet of vehicles powered by gasoline and diesel keeps increasing and adding to the **pollution** produced in many less developed countries. It can be said that the use of fossil fuels (coal, oil, and gas) is necessary for the development of society as we know it.

RESOURCES AND RESERVES

Resources are the known deposits of minerals and fossil fuels. **Reserves** are the estimate of the resources that can be extracted with current technology.

Because of its ease of extraction and versatility, oil continues to be a vital form of energy for development.

FOSSIL FUELS

Oil, coal, and natural gas are fossil fuels; that is, they are the remains of living organisms that have experienced a special process of decomposition and transformation underground.

Presently, the quantity of fossil fuels burned in one year is the equivalent of a million years of work on the part of nature to produce them and store them up.

A COMPARISON OF ENERGY SOURCES

Source		Pollution	Reserves
Nonrenewable	Coal	Yes	Limited
	Oil	Yes	Limited
	Natural gas	Yes	Limited
	Nuclear energy	Yes	Limited
Renewable	Wind	No	Limitless
	Solar energy	No	Limitless
	Energy from the sea	No	Limitless
	Geothermal energy	No	Limitless
	Biomass	Depending on use	Renewable

HOW COAL IS FORMED

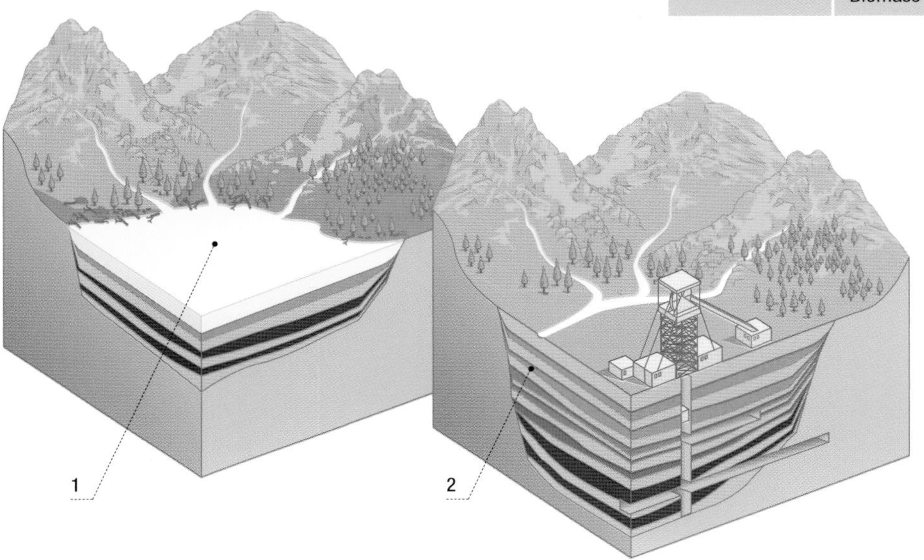

(1) Vegetation dies and remains buried in a swamp, where it changes into coal over a period of millions of years; (2) Later on new layers are formed; the best quality coal is located at the greatest depths.

It is estimated that the ecological cost of coal and oil is three times higher than the cost of their extraction and distribution. In the case of natural gas, since it is less polluting, this relationship is just double.

EOLIC ENERGY

This is the energy obtained by exploiting the force of the wind. It has been used since antiquity in the classic **windmills**, which function with huge blades on an axle much like a fan. When the wind pushes against these blades, they begin to spin on the central axis. This axle is connected to a series of gears and mechanisms that eventually connect to a piston that goes up and down and grinds the grain to produce flour. This mechanical principle is the same one used to produce electrical energy in **windfarms**. Instead of a piston for grinding grain, it is connected to a type of generator that produces **electricity**.

The main drawback to today's windfarms is that they are located on the routes of migratory birds and their blades kill hundreds of individual birds every year.

One of the main virtues of alternative energy sources is that they are clean for the environment. Another one that is just as important is that they are inexhaustible, because the Sun will not burn out and the wind will not cease to blow, and water will not stop running from a higher place to a lower one.

The energy provided by the wind is also used for certain sports.

Using natural energy sources is no modern invention.

SOLAR ENERGY

PHOTOVOLTAIC SOLAR PANELS

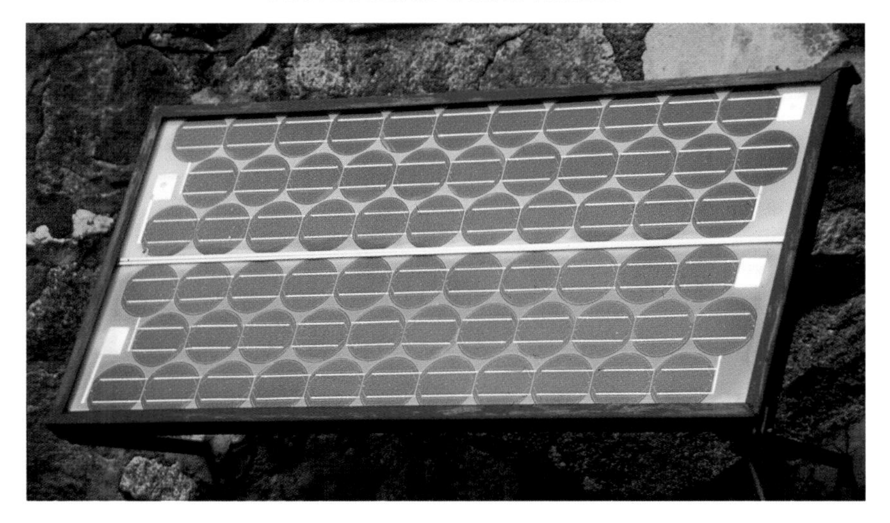

The Sun is the most important source of energy for all life forms on the planet. It is the energy that plants capture for growing, and thus, it is the basis for all ecosystems. However, its use as an energy source for humans is still very small in comparison with other types of energy. There are different techniques for collecting solar energy, and new and more efficient ways are continually being developed.

If we could exploit just 5% of the solar energy that reaches the Earth in one year, we would have the same amount of energy as in all the fossil fuel reserves.

In areas in low or middle latitudes, solar energy will soon be a feasible and affordable reality.

A multitude of small devices, such as calculators and watches, already function on photovoltaic solar energy.

SOLAR PANELS

Photovoltaic solar panels have a special coating that is capable of transforming a ray of sunlight directly into electricity. **Thermal** solar panels use the Sun's rays to heat water in a circuit; it is used for heating water for bathing and so forth.

BIODIESEL

Plant and animal food oils can be recycled cheaply and ecologically by converting them to a fuel usable by automobiles that run with a diesel motor. Also, plant wastes from agriculture can be used to produce fuel. All these forms of fuel are referred to as **biodiesel**. Thus, it is not a **renewable energy source** in the true sense of the word, but instead is a very important ecological alternative in avoiding spills that pollute water and the accumulation of wastes that are burned and contaminate the air.

Advantages of Biodiesel
• It is the only alternative fuel that works in any conventional diesel engine without any type of modification.
• It can be used pure or mixed in any proportion with diesel fuel from petroleum.
• The biological production cycle and the use of biodiesel reduce carbon dioxide emissions by about 80%, and sulfur dioxide by nearly a 100%.
• In comparison with diesel fuel from petroleum, it offers significant reductions in the output of particles and carbon monoxide.
• Various studies in the United States have shown that biodiesel reduces the risks of getting cancer by about 90%.
• It contains 11% oxygen by weight and contains no sulfur. The use of biodiesel can prolong the useful life of motors, because it has better lubricating qualities than diesel fuel from petroleum.

Many countries are already using agricultural wastes to produce biodiesel and alcohol for fuel.

Biodiesel manufacturing plants already are operating in Europe, but the attitude of U.S. industry has been more conservative.

Restaurants and other establishments that use large quantities of oils and fats must not dump the used product into the urban sewer system, but instead keep it in containers to be picked up.

BIOMASS

From the energy viewpoint, **biomass** is considered to be all the material of plant or animal origin that can be used as **fuel**. Thus, **biodiesel** is included in this definition, but biomass also included such products as wood. It still constitutes the main energy source in many undeveloped societies. Transformed into biodiesel it can reduce **air pollution**, but when it is burned as firewood it can be a major source of pollution.

Firewood is a major source of air pollution.

In poorer countries, wood constitutes nearly 90% of the fuel used.

PRODUCING ENERGY FROM BIOMASS

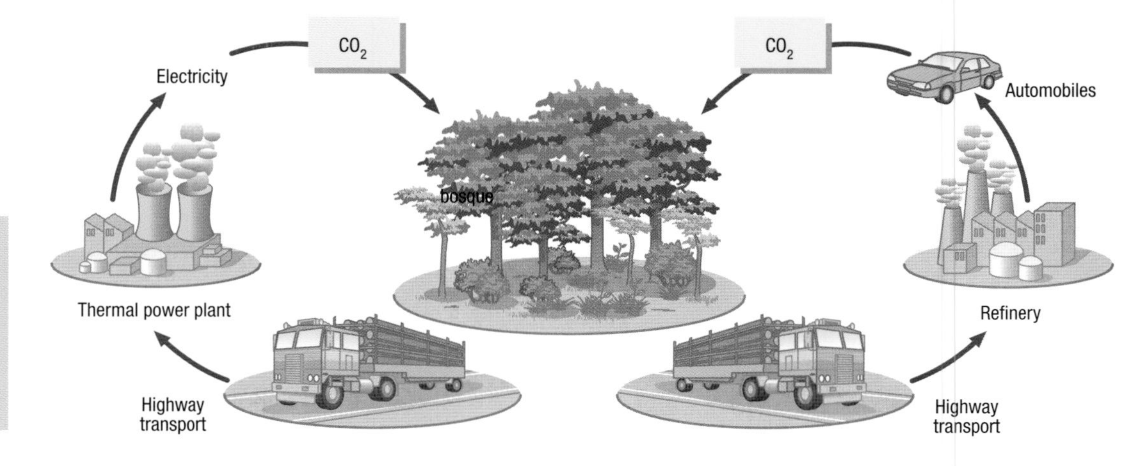

ELECTRICAL ENERGY

Electricity per se is a very clean type of energy, with no waste to dirty the environment. The problem may stem from the way this energy is produced. When it is produced by **solar panels** or windfarms, then we can speak of a totally clean energy source, because it produces no pollution at all. However, electricity currently is being produced in **nuclear** and **thermal power plants** where nuclear and fossil fuels are being burned to transform the heat produced into electricity. This process involves a fairly significant energy loss. **Hydroelectric** power plants also produce clean energy, but their presence has a major impact on the river ecosystems.

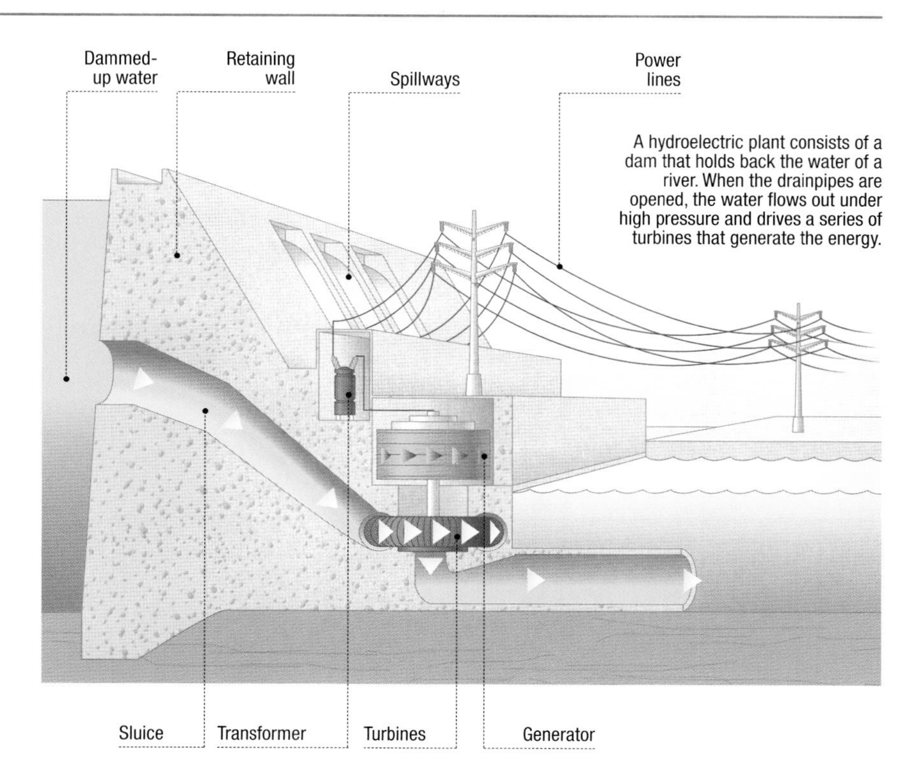

A hydroelectric plant consists of a dam that holds back the water of a river. When the drainpipes are opened, the water flows out under high pressure and drives a series of turbines that generate the energy.

To avoid interfering with the travels of migratory fish, some dams are equipped with special fish ladders.

Changes in the water level of dammed-up rivers cause serious changes to the ecosystem. Many species disappear.

Like windmills, water mills were invented centuries ago; they use the force of the water to perform mechanical work.

WHY ARE ALTERNATIVE ENERGY SOURCES USED SO INFREQUENTLY?

At the international summit in 2002 in Johannesburg, South Africa, pressure from the oil industry managed to keep delegates from passing resolutions that would reduce pollution.

This is a question that anyone would ask after learning about their characteristics. The main reason is that they are still expensive to produce, and the energy generated is often inadequate for satisfying the present high **consumption**. In some countries the government offers aid and subsidies for promoting the use of these energy sources, and the current tendency is toward increasing use. The main opposition to these clean energy sources is the **oil industry**, which considers them a threat for its economic interests, so it boycotts the international meetings intended to promote them. However, nonrenewable resources will become increasingly expensive, and it will be more cost effective and essential to use renewable resources.

A fifth of the world's population consumes two thirds of the planet's energy. If all citizens of underdeveloped countries reached the standard of living of the West in 25 years, the present consumption of energy would increase five times.

RECYCLING IS SAVING

In nature, nothing is thrown away or wasted. If we look at the **cycles of matter** and energy or the **trophic chains**, we see that there is always one organism that exploits what another leaves behind, and when it dies, it becomes food for another organism. Humans are accustomed to use and discard, continually producing more **waste**, which may eventually bury us. The only solution is to imitate nature – that is, to **recycle** both materials and energy, and thereby conserve resources, which in many cases are scarce and irreplaceable.

WHAT IS RECYCLING?

Recycling means returning something to a **cycle**, and because we are speaking of all the **materials** that we use in our daily lives, this involves reintroducing them into the production cycle—in other words, not throwing them away, which would be to remove them from the cycle. Recycling a product means using the material or the components from which it is made to produce a different product. That way we avoid having to use new material, which may be in short supply. Thus, recycling is a way of **saving**.

The containers are a way to sort discarded materials at the outset.

RECYCLABLE PRODUCTS AND MATERIALS

A small sampling of the things we use in the home and that can be recycled.

ALMOST EVERYTHING CAN BE RECYCLED

In the following chart some lines are left blank. Fill them in yourself; first think of things that you find around you and that you use in your daily life.

Material or produce	Recyclable	End Products	Advantages
Paper, cardboard	Yes	Paper, cardboard	Forests are not destroyed.
Plastic containers	Yes	Plastic items	No pollution, petroleum savings
Glass	Yes	Glass products, road surfaces	No pollution, energy savings
Batteries	Yes	Minerals, plastics	No very dangerous pollution
Automobiles	Yes	Metals, plastics	No pollution, raw materials savings
Water	Yes	Water for drinking, irrigation, industry	Water is saved and purified.
Organic waste	Yes	Compost, fertilizer	Soil preservation, no spills
Waste from nuclear power plants	No	None	None
Computers, stereos	Yes	glass, metals	No pollution, raw materials savings
Natural fabrics	Yes	Paper, cardboard	Forests are preserved.
Synthetic fabrics	Yes	Industrial plastic	Pollution is avoided.
Wood items	Yes	Plywood, paper pulp	Forests are preserved.
Agricultural wastes	Yes	Fertilizers, biodiesel	Oil is conserved.

RECYCLING AT HOME

Recycling is not so difficult to do. We are going to make some **recycled paper**. The result will not be perfect because we do not have everything we need, but this will show you that it is possible:

1. Tear several pages of newspaper into pieces and put them into a bucket with water.
2. Set aside for two or three days.
3. Grind up the resulting paste.
4. Heat the paste and a little detergent.
5. Get rid of the ink that comes to the top, along with the foam.
6. Let it cool down and strain it.
7. Spread out a sheet of paste about 3/32" (2–3 mm) deep on a rectangular piece of screen.
8. Put sheets of newspaper on top and weight it down.
9. Wait several hours and remove the sheet of paper.
10. Let it dry.

INTERESTING FACTS

- Nearly 40% of the paper that we use is recycled.
- To produce the material for one aluminum can, a half-can of oil is required.
- In some countries, nearly 80% of the steel produced is made from scrap metal.

A SIMPLE WAY OF MAKING PAPER

1, 2 3 4, 5

6, 7

8, 9

10

BEFORE RECYCLING WE HAVE TO MAKE SOME CHOICES

RECYCLING AN AUTOMOBILE

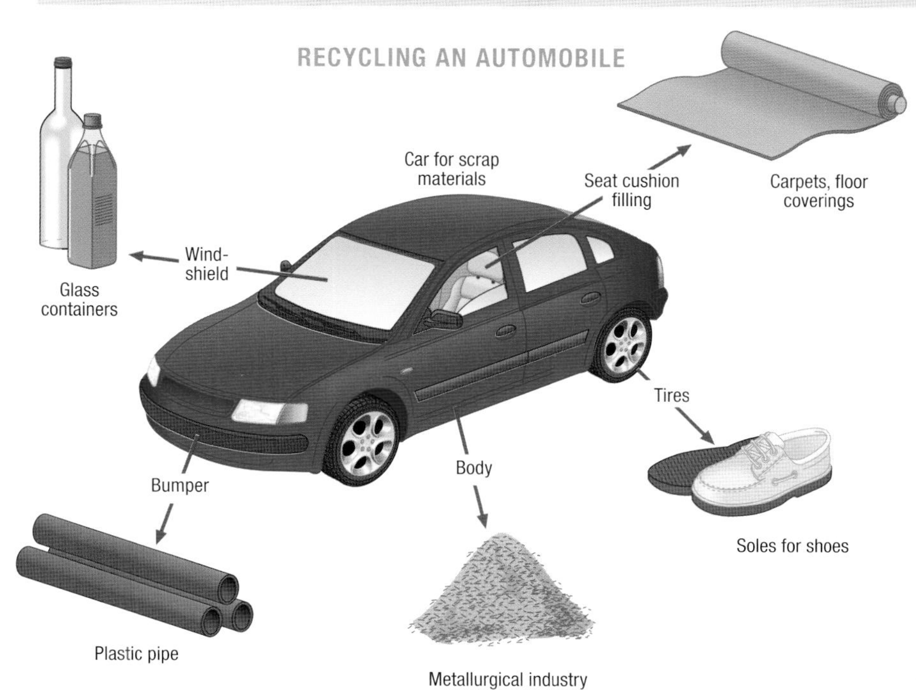

To make use of **wastes** we first have to sort them. **Recycling** companies have to do this, but we too can contribute to the effort by sorting our own trash. For that purpose there are containers where we can deposit glass, others for plastics, others for paper, and so forth. That way the sorting that takes place in a **trash treatment plant** will be easier. Materials of the same type are crushed up or baled and sent to factories that use them as raw materials.

A goal for automobile manufacturers is for more than 90% of the vehicle to be recyclable.

DESERTIFICATION AND SOIL MANAGEMENT

The **soil** is not just ground-up rock; it also contains a variable amount of **organic** matter, and this makes it **fertile**. Fertile soil is not available everywhere on Earth, and that is why agricultural activity is limited to areas that have such soil.

In recent times, because human activity, there is an ongoing process of fertile soil loss leading to increasing desertification on the planet.

SOIL

In some areas of the planet the thin layer of the Earth's surface forms a special structure known as **soil**. On the one hand this is the product of chemical and mechanical activity, which reduces the **bedrock** and forms small stones, and fairly fine gravel and sand. On the other hand, the soil is also the product of **biological activity** by the life forms that inhabit it.

Organic matter is present in the form of plant remains (sticks, leaves, fruits, and the products of their decomposition) and wastes of animal origin (droppings, dead bodies, and so forth). In addition it contains millions of **microorganisms** (bacteria and protozoa, for example) and other life forms (insects, worms, fungi, etc.).

Presently a bit more than 11% of the planet's surface is suitable for crops but that could be increased to 24%.

The soil is constantly exposed to chemical, physical, and biological activity. The photograph shows a beech stand in the fall.

PERCENTAGES OF SOIL USES

Continent	Crops	Pasture		Other
Africa	6	26	24	44
Central and North America	13	16	32	39
South America	7	26	54	13
Central and Northern Asia	10	21	32	37
South Asia	24	21	13	42
Southeast Asia	17	5	57	21
Australia	6	55	18	21
Europe	30	18	32	20

Organic material makes up just 1 to 2% of the weight of soil, but it is the essential part for its fertility.

These are approximate percentages because the use of the soil in recent years is experiencing great changes, mainly a reduction in forest areas.

HUMUS

This is the portion of the soil's organic material produced by the decomposition of organic remains. It is the main factor in determining its fertility.

EROSION

This is the natural process by which **weather phenomena**, such as wind, water, and ice, wear away the surface of the planet. Erosion causes very old mountains to have peaks and shapes that are more rounded and smoother than more recent mountains. Erosion also acts on the soil, eliminating or reducing its thickness. However, **vegetation** (mainly the forests) acts like a protective layer and prevents that loss. Human activities encourage erosion by directly eliminating the soil (**urbanization**), exploiting it excessively (agricultural and livestock **overexploitation**), or eliminating the protective layer (**cutting** down forests).

TERRACED AGRICULTURE

This is a system that divides the slope of a mountain into small steps to prevent erosion.

The formation of one inch (2.5 cm) of soil can take between 100 and 2,500 years. It can take less than one minute to destroy an inch of soil.

Cutting a forest for use in industry (construction, paper, etc.) should be followed by reforestation. The photograph shows a cut forest on Vancouver Island, Canada.

THE ADVANCE OF THE DESERT

Continent	Locations	Causes
Africa	Northeast	Erosion
	Sahel	Grazing, agriculture
	Botswana	Grazing
America	Central U.S.	Agriculture, livestock
	Central Mexico	Erosion, drought
	Northeast Brazil	Urbanization
Asia	Middle East	Erosion
	Central Asia	Grazing, irrigation
	Mongolia	Grazing
	Yangtze, China	Agriculture, urbanization
	Southeast Asia	Deforestation, erosion
Australia	Southeast	Agriculture, livestock
Europe	Southeastern Iberian Peninsula	Urbanization, agriculture

DESERTIFICATION

This is a process by which fertile ground becomes sterile and incapable of sustaining plant life.

Clearcutting without reforestation can lead to desertification of the terrain.

SOIL MANAGEMENT

To avoid the effects of **erosion** and **desertification**, it is necessary to treat the earth in a way that avoids or minimizes damage and encourages its regeneration. This is what is meant by **soil management**. In agriculture, **crop rotation** is practiced; this allows the cultivated land to recover its fertility by having different plants in different years. The planting of **hedges and forests** helps avoid the effects of erosion. Controlling **grazing** is also very efficient in avoiding the disappearance of the plant covering in sensitive areas such as the dry savannas and prairies.

Often it is a good idea to surround cultivated fields with hedges and small forests to protect them from livestock and to encourage wildlife.

Three percent of cultivatable land is in serious danger of desertification, and 12% is in elevated danger.

OVERFISHING AND OCEAN MANAGEMENT

We have already seen that in regard to pollution the oceans are not a drain into which we can keep pouring all the wastes of our civilization. The same is true of the ocean's resources. They too can become exhausted. Some, such as the minerals deposited on the bottom and fossil fuel deposits located under the oceans, will be used up in the same way as the resources on terra firma. Others, such as the energy of the seas and ocean currents, are inexhaustible or renewable, like fishing.

Cephalopods are one of the most overfished species.

OVERFISHING

Traditional fishermen have managed to produce their products throughout the centuries without altering the **population** of marine animals. Still, starting at the start of the twentieth century, new techniques have made it possible to construct tremendous **freezer ships**, where the fish caught continually by hundreds of auxiliary fishing boats are processed immediately. The result is that the fish populations have not been able to recover, and many have disappeared. This excess exploitation is known as **overfishing**.

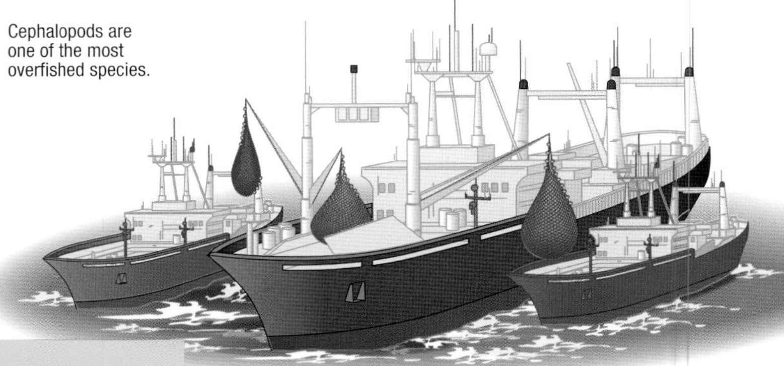

Trawlers drag fishing nets along the bottom and not only catch species of commercial interest but also destroy the whole ecosystem.

EXPLOITATION OF THE OCEAN'S LIVING RESOURCES

Type	Example
Plants	Edible seaweed, seaweed for animal fodder, seaweed for industrial uses
Mammals	Whales, seals
Bottom fish	Plaice, halibut, hake, ray, cod
Surface fish	Tuna, sardines, anchovies, herring, mackerel
Crustaceans	Lobster, prawn, crabs
Mollusks	Mussels, oysters, clams
Cephalopods	Octopus, squid, cuttlefish

DRIFT NETS

These are huge nets several kilometers long that float on the ocean and catch all kinds of marine animals in addition to edible species. Turtles, dolphins, and even ocean birds get caught in the nets and die.

A visit to a fish market will give you an idea of the ocean's richness, although it is becoming more of a luxury to eat fresh fish.

USES FOR FISH

Type	%
Fresh or frozen fish	35
Feed and oil production	32
Canned fish	16
Smoked, salted fish, and others	17

THE DISTRIBUTION OF THE OCEAN'S RICHES

It is possible to distinguish main zones in the oceans. One is the **littoral**, or coastal region, which surrounds the continents and is made up of the **continental shelf** (down to a depth of about 620 feet/200 meters). The other is the **high sea** (the **pelagic** region). **Plant plankton** grows principally near the continents, which is why the continental shelf is the area where most marine life is concentrated, and this is where the main **fishing grounds** are located. Mineral riches, on the other hand, are found mainly in the depths of the ocean, that is, in the central areas of the oceans. But oil resources are found in both regions, although the deepwater deposits cannot yet be exploited because of technical difficulties.

THE ZONES OF THE OCEANS

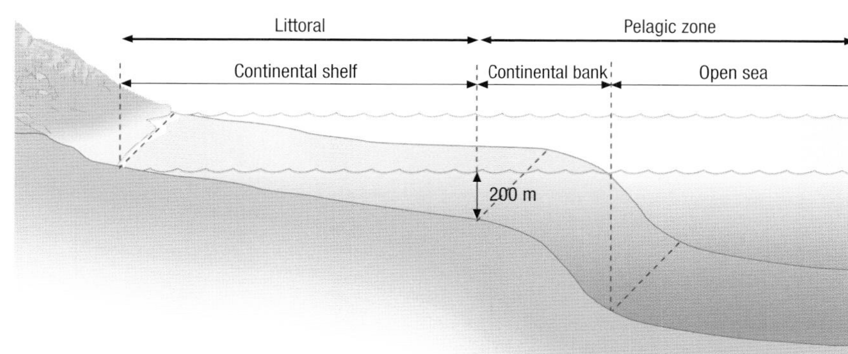

Minerals from seawater are produced in extraction plants located on the coast.

It is believed that the greatest coal deposits on the planet are located beneath the Trans-Antarctic Mountains.

SOME MINERAL RESOURCES IN THE SEA

Resource	Location
Petroleum, natural gas	North Sea, Gulf of Mexico, coast of Ecuador, coast of Venezuela, the coasts of Indonesia, the Red Sea, the west equatorial coast of Africa
Manganese nodules	North Atlantic, South Atlantic, central Indian, North Pacific, South Pacific
Silver, zinc	Red Sea
Mineral-rich sediments	Equatorial Atlantic, North Indian, central Pacific
Uranium	Saltwater

Half of the world's petroleum and natural gas reserves are located in the continental shelf. The photograph shows offshore oil drilling in the North Sea.

MANAGING THE OCEAN'S RICHES

Part of the biological and mineral riches of the oceans is located within the territorial waters of various nations, but another significant part is in international waters. This makes it necessary to establish **treaties** that regulate the use of these resources. Because the oceans are an environment shared by the entire planet, international cooperation is necessary for their management, as in the case of the **Antarctic Treaty**. There are numerous mineral and fishing resources on that continent, although the **Antarctica Moratorium** of 1991 established a period of 50 years to begin exploitation, thus providing time for the development of technologies that will not destroy this valuable **ecosystem**.

These elephant seals on Livingstone Island (Antarctica) can enjoy their peace for the moment, because of the moratorium that prohibits exploiting the riches of this continent.

BIODIVERSITY AND THE EXTINCTION OF SPECIES

The millions of years that have gone by since the appearance of life on our planet have given rise to a tremendous variety among organisms in the course of evolution, which is what we call biological diversity or **biodiversity**. This variety is the essential basis for the continuance of life on earth. Human activity and its negative effects have accelerated in recent decades, and a great number of **species** have become extinct before evolution was able to function, and that is a great loss for the planet.

THE RICHNESS OF LIFE

Life appeared in the ocean more than 3 to 3.5 billion years ago, and it did so in the form of tiny **one-celled animals**. Later on there were creatures made up of several cells, the **pluricellular animals**, which experienced great expansion. Because the planet was empty, they could take on the most extravagant forms, but as their numbers increased, **competition** began, and the only survivors were the ones that could adapt to the environment in which they lived. Since then evolution has produced millions of forms. Many of them have become extinct (such as the **dinosaurs** and giant ferns), and others have arisen (such as modern birds and mammals). But for evolution to continue function, there must be a great number of species. That is the only way to guarantee that there will be enough material for it to work.

There is a tremendous variety of different species in the coral reefs.

Between 1.7 and 1.9 million species are known, but scientists calculate that some 5 to 30 million species remain to be discovered, including some mammals.

Every year 20 new reptile species are discovered.

KNOWN SPECIES (rough estimates)

Plants	Lower	100,000
	Higher (vascular)	250,000
Animals	Invertebrates	1,350,000
	Vertebrates	50,000

Despite their protection in national parks, some species are in danger of extinction. Above, the Amazon parrot, a rare bird about 12 inches (30 cm) long; below, the blue-faced monkey of Costa Rica.

THE IMPORTANCE OF GENETIC RESOURCES

Genes are the carriers of the key to life. There are a great variety of genes on Earth, and each one gives a plant or an animal a specific shape or ability to survive in certain conditions. If most of the plants or animals disappear, these possibilities disappear with them; thus, if the planet's environmental conditions change and there is no gene capable of adapting to them, **life** could disappear. This is the importance of the reserve of genetic resources. And this **reserve of genetic resources** is based on the existence of very different organisms—that is, ones that exhibit **great biodiversity**.

The Amazon jungle is one of the most diverse terrestrial ecosystems on the planet. There are thousands of species that live there and that science still has not discovered. This represents a tremendous genetic reserve, which is why it is essential to preserve this ecosystem. The photograph shows the rare *hermit ibis.*

NATURAL EXTINCTION

In the course of evolution there are species that cease to be adapted to changes in the environment in which they live, or which are less suited than other more recent species that compete with them and win out. The old species ends up becoming extinct. This is a natural process that has been repeated thousands of times throughout the **history of life**. The huge **dinosaurs** dominated the Earth for some 130 million years. Then, perhaps gradually, they disappeared in a few thousand years.

THE EXTINCTION OF LIFE IN A LAKE BECAUSE OF ACID RAIN

Lake condition	pH	Consequence
Healthy	7	Great diversity of life
Slightly affected	6	Disappearance of all crabs and some fish
Strongly impacted	5	Nearly all fish disappear
Catastrophic situation	4.5	A few eels survive
Final state	4	Scarcely a few bacteria survive

The pH indicates the water's degree of acidity. A value of 7 indicates neutral water, and the lower the number, the more acidic the water.

Great climatic conditions produced over hundreds of thousands or millions of years have displaced many animal and plant species from their original habitat or else have caused them to become extinct.

Mountain gorillas, which live in the misty mountains around Lake Kivy (between the Democratic Republic of Congo and Rwanda) are on the road to extinction because of poaching.

ARTIFICIAL EXTINCTION

Humans are the only known animals that are capable of intervening in **evolution**, selecting some species over others, depending on their interests. However, there are many details that we do not know about how evolution works, so this activity is very dangerous, including for our own survival. Humans have caused the disappearance of many **species**, in some cases by killing them directly, and in others by destroying their habitat or their way of life. With these artificial extinctions, life has not created a form to take the place of the one that disappeared (as it happens in natural extinction); as a result, there is a "hole" in the network that all of us life forms make up. Many holes may break up the network entirely.

It is now estimated that some 60 to 100 species become extinct every day.

The disappearance of one species implies the disappearance of others that depend on it; if bees were to disappear, many plants would become extinct, because they depend on bees for pollination.

THE DEVELOPMENT OF HUMANITY

The **human species** is part of the rest of the species that inhabit the Earth. It is subject to the same physical laws and the same **evolutionary** mechanisms as the other organisms, although it has developed a special ability to intervene in **natural processes**. As a result, the development of humanity is of great importance for the survival of **life** on our planet.

THE CONQUEST OF THE PLANET

The first human inhabitants lived in small, isolated groups. Just like other land-based **primates**, they fell prey to many predators that were stronger than they were. Still, the increasing intelligence level of the species made it possible to survive under adverse conditions. Since that time the **growth** of the human population has been constant. This has led to the occupation of most of the **habitats** that are suitable for life. Humankind's increasing independence from natural enemies means that people now are present all over the planet, exerting increasing pressure on the **ecosystems** and other inhabitants of the Earth.

THE HUMAN PLAGUE

When one species ceases to fall under the **control** of its natural enemies it becomes a **plague**, as with the rabbits in Australia. We humans scarcely have any natural enemies, so we have become a plague.

The population of the world is distributed unevenly, the same way riches are.

SOME CONSEQUENCES OF HUMAN OVERPOPULATION

Necessary resource	Solution adopted	Effect on nature
Space	Urbanization	Destruction of natural habitats Reduction of wild animal range Disappearances of species Waste generation
Water	Exploitation of Aquifers and swamps	Disappearance of wetlands Desertification Extinction of aquatic flora and fauna
Food	Agriculture, livestock	Destruction of natural habitats Reduction of wild animal range Waste generation Deforestation Desertification Extinction of species Pollution
Transportation	Automobiles, airplanes, ships	Air pollution Water pollution Noise pollution Reduction of wild animal range
Energy	Thermal, nuclear, and hydroelectric power power plants	Air pollution Water pollution Dangerous wastes
Goods	Factories	Air pollution Water pollution Soil contamination Deforestation

India is a country of great contrasts. While a minority live in luxury, the vast majority of people live in poverty.

THE THIRD WORLD

This is the term used to designate the poorest countries on the planet, which include more people than the rich countries, which are referred to as the First World.

OVERPOPULATION

Changes in ways of living and scientific advances have been so rapid that **traditional societies** have not had time to adapt. The large families of past centuries were necessary to compensate for the many deaths; the number of children who reached adulthood for every couple was reduced, and population grew very slowly. **Medical progress** now makes it possible for many children to survive, so that the number of children who reach adult age per couple is far higher than scarcely a century ago. The consequence is a dizzying increase in **population**.

In poor countries, the lack of infrastructure means that the soil and water are highly polluted.

ZERO GROWTH

When the number of residents in a country who die is equal to the number of births, it is said that the country has zero growth.

Almost 15 million children less than five years old die every year from starvation and disease. The photograph shows a mother and her child in Jaipur, India.

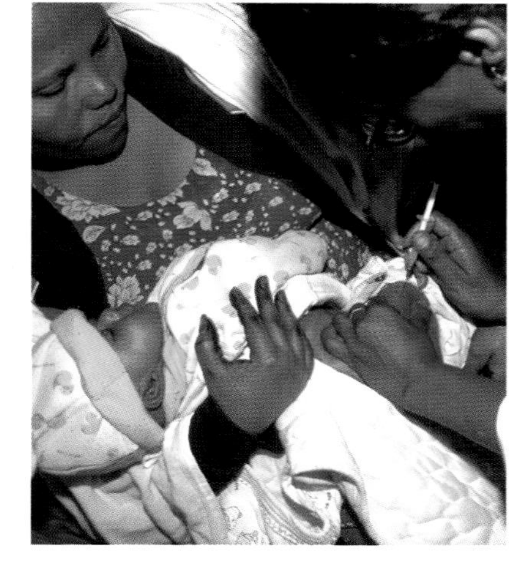

The current population of the planet is around 6.2 billion people.

In developing countries, the population increased by 2 billion to 4 billion in just 30 years.

Even though vaccination campaigns prevent death by disease of millions of children in the Third World, a lack of food in those places condemns them to death by starvation.

SOME SOLUTIONS TO OVERPOPULATION

INTERNATIONAL ORGANIZATIONS

The UN is the United Nations, the FAO is the Food and Agriculture Organization, and the WHO is the World Health Organization. These are three of the many organizations that are concerned with human development.

Excess human **population** is a serious problem, especially in developing countries. The resources necessary for development have to be used for feeding an increasing population, which also increases **poverty** and pressure on natural **ecosystems**. International organizations such as the **UN**, the **FAO**, and the **WHO** have proposed various plans to limit this excess population. Among the measures proposed (which are already being applied in many countries) is **birth control** but that has to be accompanied by education so that people understand the necessity, and development aid so that families do not keep having children just to supply the work hands they need to survive.

The world headquarters of the World Health Organization (WHO) in Geneva, Switzerland.

FOOD FOR HUMANS

Humans are **omnivores**; that is, we eat both plants and animals, so we use many of the **resources** that nature offers us. **Agriculture**, **livestock raising**, and **fishing** are essential activities for our lives. The present **overpopulation** creates a huge and destructive impact on natural environments.

FROM NOMADIC HUNTER TO FARMER

Prehistoric people were **nomadic hunters** and **gatherers** who depended on the way of life of their prey and the seasons in which the plants they used for food grew. Some isolated tribes still follow that way of life (e.g., in the Amazon region). Civilization dawned when people began raising their food animals instead of hunting and raised the plants they needed. That was the birth of **agriculture** and **livestock** raising. People's food supply was thus more guaranteed, but that also made population growth possible.

SUBSISTENCE ECONOMY

This term refers to the way of life that allows an individual just to survive with minimal resources.

AGRICULTURE

Agriculture was born in the fertile areas along the large rivers, apparently in **Mesopotamia**. At first the most common edible plants were grown. Later on the most productive ones were selected, and today we have **varieties** that produce a higher yield. The first farmers did all the work by hand, but later they began using draft animals. The agricultural revolution arrived along with **mechanization** (tractors, harvesters, etc.), but it left many farmers unemployed, because the new techniques required far less labor.

In underdeveloped countries, agriculture is not very mechanized because labor is so cheap. The photograph shows a plantation in Rajastan, India.

In industrialized countries people eat more than 100% of what they need; the result is an increase in obesity. In poorer countries people eat less than 85% of what they need; this leads to malnutrition.

MALNUTRITION

Malnutrition occurs when an adult has less than 1,500 calories per day.

Mechanization in the field has the drawback of almost entirely eliminating hand labor. In a single day, for example, one tractor can do the work of 100 people in a week.

BASIC CROPS

Species	Use	Growing area
Wheat	Staple food, for making bread. Contains 8–15% protein	Temperate climates
Rice	Staple food in Asia. Contains 8–9% protein	Tropical climates, in waterlogged areas
Corn	Staple food in Central America and Africa; contains 10% protein. Also used for livestock	Hot and temperate climates
Potato	Staple source of carbohydrates in many countries	Cool and temperate climates
Barley	Mainly for forage and making beer	Cool and temperate climates
Sweet potato	Secondary food rich in starch	Moist, tropical climates
Soy	Contains between 30 and 50% Protein; staple food in many poor countries.	Warm climates
Tapioca	Low protein content; resistant to drought; staple food in Africa	Hot regions
Rye	For making bread	Cool, moist climates
Oats	Livestock fodder	Cool, moist climates
Sorghum	Staple food in dry regions of Africa and Asia	Warm, dry climates

LIVESTOCK

The process of **domesticating** animals paralleled the development of agriculture. Humans selected gentle **herbivores** and bred increasingly productive strains. Domestic **livestock** is used as work animals and for producing meat, milk, and hides. Nowadays many farms use industrial procedures in handling these animals.

Wool is the processed hair of sheep; it has been used since antiquity. Currently, it is experiencing competition from synthetic fibers.

Animal protein is more expensive to produce than producing the same quantity of plant protein.

Between 40 and 75% of the grain production in many countries goes to feeding livestock.

In many countries the protections and subsidies accorded to livestock operations lead to overproduction (above), yet in other places on Earth people can scarcely subsist (below, a woman in India milking a water buffalo.

USES FOR DOMESTIC LIVESTOCK

Species	Uses
Cow, bull	Meat, leather, milk, work
Zebu	Meat, milk, work
Pig	Meat, skin
Sheep	Meat, wool, milk
Goat	Meat, leather, milk
Horse	Work, leisure, meat
Camel	Work, transportation, meat, milk, leather
Reindeer	Work, transportation, meat, milk, leather
Yak	Transportation, meat, milk, leather
Chicken	Meat, eggs
Guinea fowl	Meat
Ostrich	Meat, feathers
Goose	Meat, eggs, down
Duck	Meat, eggs

ECOLOGICAL BEHAVIOR

Ecology is a science that attempts to explain how nature works and to provide ideas for preserving it; but we are the ones who have to put these ideas into practice. If the private citizens do not behave in a harmless way with respect to the environment, the work of scientists comes to naught, and furthermore, their conduct causes harm to everyone else on the **planet**.

WATER USAGE

Freshwater represents just 3% of all the water in existence on the planet. But in addition, of all freshwater, the majority is in the form of ice, so we can use less than 1% of it. This is thus a very scarce commodity that needs to be managed. **Recycling**, reducing losses though **leaks**, irrigation techniques that avoid waste, and growing plants that require little or no watering are among the principal measures for saving water.

Various daily situations in which we can save water through thoughtful use.

To maintain a reasonable quality of life, some 22 gallons of water per person per day are required.

Saving	Wasting
Washing with a flow regulator on the faucet	Washing with the faucet wide open
Showering	Taking a bath
Watering the garden with a trickle irrigation system	Watering with a hose
Shutting off faucets tightly	Letting faucets drip
Growing water-sparing plants	Growing water-thirsty plants
Doing a full load of laundry	Doing a partial load of laundry

There are great differences in domestic water consumption between poor nations and rich ones; the difference can be up to 100 times.

OLD BATTERIES

It is very important to place old batteries in designated containers, because they contain many pollutants and substances that are harmful to the environment.

CHART SHOWING HOW BATTERIES ARE RECYCLED

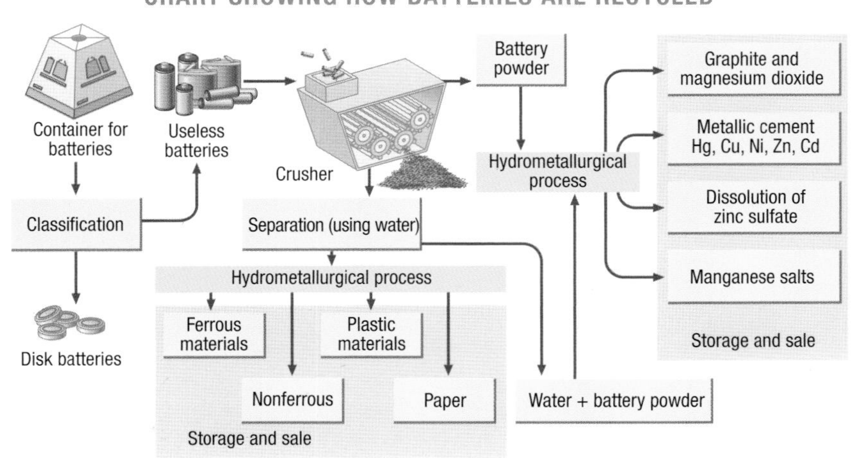

HOUSEHOLD TRASH

One of the major problems for modern society is getting rid of the **waste** it produces. The average citizen can contribute to a solution for this problem by using the available means for getting rid of household wastes. **Plastic** and **glass** containers, **paper** and **cardboard**, and **organic wastes** must be sorted into different bags and each one deposited into different containers. That way, every type of waste is put to a different purpose (**recycling**, composting, etc.). Further, used-up batteries, expired **medicines**, and **electronic devices** should not be thrown into the trash. These are measures that avoid polluting the environment.

HOW TO ENJOY NATURE PROPERLY

Surely you like to find the land clean when you go out on a field trip. That is true for everybody, including the wild animals. For that to happen, we all have to be respectful of nature. There are a few basic rules that we must always follow: discard no **trash**, cause no **fires**, avoid polluting **rivers** (e.g., as in washing a car), pull up no plants, make no **noise** (animals and people need quiet), do not bother the animals (if you frighten a nesting bird, it may abandon the eggs and the chicks will die before they are even born).

When you visit a protected area, you must strictly obey the rules posted at the entrance.

On an outing we must respect the plants, animals, crops, fences, and so forth. We leave the land in the condition we would hope to find it.

MEANS OF TRANSPORTATION

Type	Advantages	Disadvantages
Airplane	Speed over long distances	High fuel consumption, pollution
Train	Speed over short and medium distances	Few
Private auto	Mobility	Pollution
Shared auto	Savings per person, reduction in traffic and pollution	Lacks of flexibility; also, pollution
Sailboat	Silence, no pollution	Slow
Motorboat	Faster than a sailboat	Noise, pollution
Walking	Flexible within city, good for Health	Fatigue

TRAFFIC

If drivers followed certain simple rules, not only would they produce less pollution, but they would also save fuel, which is becoming more expensive every day: drive moderately, avoid rapid acceleration, keep **tires properly inflated**, keep the **engine** tuned properly, avoid carrying cargo on the roof if not necessary, and avoid overloading (which is also dangerous). In choosing a vehicle, search for the one with the best fuel economy (which also reduces **exhaust emissions**) and that runs on the least polluting fuel.

Whenever possible use public transportation. That way you will help cut down on traffic and pollution.

THE GOOD CONSUMER

You can help improve the **environment** by choosing how you shop and the type of purchases you make. In addition, you will save money. Prepare a list of what you want before going into a store. That will keep you from buying things you do not need. Check the labels to see if the product you are looking for contains substances that are harmful to the environment, and if so, put it back on the shelf. Buy only wood products that contain **certified wood**. Avoid excess packaging. If you go to the supermarket try to buy natural products and avoid highly processed ones and products that contain **genetically engineered** substances or ingredients.

Ethical purchasing means demanding that price take into account the cost of not harming the environment and fair compensation to the producer.

CERTIFIED WOOD

Wood with a seal guaranteeing that it comes from controlled forest plantations.

Before shopping, have a clear idea of what you need. At point of purchase, choose products that are most respectful of nature and represent fair business practices.

THE PLANET AND THE NEW TECHNOLOGIES

The intervention of humans in **nature** has ranged from minimal in prehistoric times to massive and harmful in present times. This is linked to **industrial development**. But the growing worry of everyday people about the **environment** has opened up the way for new technologies, some of which are helpful in avoiding the damage done and even repairing it.

COMMUNICATION

Nowadays everyone is in communication with everyone else, and it takes only a few seconds for news to travel around the whole planet. **Television**, **radio**, **cell phones**, the **Internet**, and modern **audiovisual media** are technical advances that have some disadvantages (mass production, manipulation of opinion, physical risks from radiation, and so forth), but they also have very significant advantages for the care of nature. These techniques and devices make it possible to find out about assaults on the environment immediately and try to correct them, raising people's awareness of **conservation**, and organizing popular movements to demand respect for nature.

Artificial satellites provide very useful information for predicting the weather.

Thanks to modern telecommunications media, any accident or assault against nature can become known and addressed.

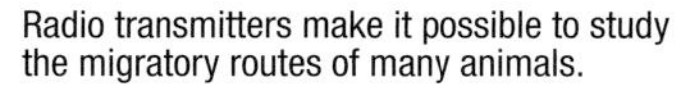

Radio transmitters make it possible to study the migratory routes of many animals.

The Internet makes it possible to access the most current information about nature from any place in the world.

GENETIC ENGINEERING: GOOD OR BAD?

A genetically engineered organism is one into which genes from a different organism have been introduced. This is the case with corn, wheat, tomatoes, and a long list of other products. These transgenic organisms have genes that allow them to continue looking good for a long time even though they have begun to lose their freshness inside.

As it happens with any technique, **genetic engineering** is not good or bad in and of itself but only in function of its ultimate use. One positive aspect is its possible application to **medicine** for curing genetic diseases that might cause the patient's death. A negative aspect is the use of **transgenic foods**, because, although they are not dangerous themselves, we do not know how these artificial organisms act, so they could trigger allergic reactions, mutations, and more. The freeing of transgenic organisms into nature is a threat to **biodiversity** and a danger to health, for they could facilitate transmission of viruses among organisms.

Farmers who grow transgenic crops have to buy new seeds from suppliers every year because the seeds produced by the plants have been engineered to be sterile.

CLONING

Cloning involves producing a new individual from a preexisting one; it thus has **identical genes** and is identical to it. This technique is widely used with **microorganisms**, but, with vertebrates, it is still in the experimental phase. In plants, on the other hand, it is a common practice. The main advantage lies in producing many individuals with favorable characteristics. In laboratory cultures it is possible to clone bacteria and other microorganisms to produce a very large population in a few hours.

HUMAN CLONING

In addition to ethical objections, the cloning of humans has not been totally perfected technically.

Dolly, the first cloned sheep in the world, experienced unusually rapid aging, and that may affect all cloned mammals. Dolly was born in July of 1996, and she died in February of 2003; her lifespan was half the usual expectation for a sheep. Today she is stuffed in the Royal Museum in Edinburgh.

MARINE CULTURES

Just as **raising livestock** replaced hunting, and **agriculture** replaced seasonal gathering, marine cultures are the present-day alternative for fishing and seaweed gathering. Nevertheless, the cultivation of aquatic organisms is very old, and the Chinese were breeding carp at the time the Roman Empire was breeding oysters. **Pisciculture** is highly developed for certain species, such as the trout in freshwater and the salmon in the sea. **Aquaculture** makes it possible to guarantee consistent production and uniform quality, while at the same time reducing pressure on natural ocean or river ecosystems.

AQUACULTURE

This is the set of techniques and procedures for raising aquatic organisms.

PISCICULTURE

This is the raising of fish in enclosed areas, where they are fed and protected from their natural enemies.

For decades fish factories have fulfilled a double mission: on the one hand, providing rivers with young fish, and on the other, providing different fish species for human consumption without exhausting the ecosystems.

PROTECTED AREAS

Human presence has progressively reduced the area occupied by **unspoiled nature**, and human activities have polluted or destroyed large areas of the planet. As a result, it has been necessary to create **protected areas** (**national parks**, **preserves**, etc.). The goal in these areas is to preserve nature and its flora and fauna in their original condition by avoiding human intervention and their destruction. They have also been converted into a **last refuge** for many species. That is why they play an important role in the continuity of **life** on Earth.

THE NATIONAL PARK OF THE GALAPAGOS ISLANDS (ECUADOR)

It was on these islands that **Darwin** got the ultimate inspiration for his **theory of evolution**. Because the islands are isolated from the mainland, the species that arrived here evolved independently and gave rise to new species adapted to the conditions of their **habitat**. There are many animal **endemisms**, that is, species that live nowhere else but here, plus some 625 plant species.

THE GALAPAGOS ISLANDS	
Location	In the Pacific Ocean on the equator, about 600 miles (1,000 km) from the coast
Geography	19 volcanic islands
Area	30,900 square miles (80,000 km^2) (land and sea areas)
Creation	1936
Ecosystems	Coastal marine, mangroves, beaches, highlands
Notable species	Cacti, mangroves, marine iguanas, land iguanas, giant tortoises, Galapagos penguins, Darwin's finch

Since the Galapagos Islands are far from the coast and off the traditional routes, they have maintained much of their original characteristics. The photograph shows the Island of San Bartholomew.

MANÚ NATIONAL PARK (PERU)

This park encompasses one of the tributaries of the **Amazon River** and is located in its upper basin. It goes from an altitude of 490 feet (150 m) up to over 13,800 feet (4,200 m). It is one of the protected areas that is most representative of the **Amazon jungle**, and it possesses amazing **biological diversity**. More than 850 bird species (15% of the entire world) live here, plus 100 species of mammals and more than a half a million species of arthropods. It contains 14 different types of **tropical forest**.

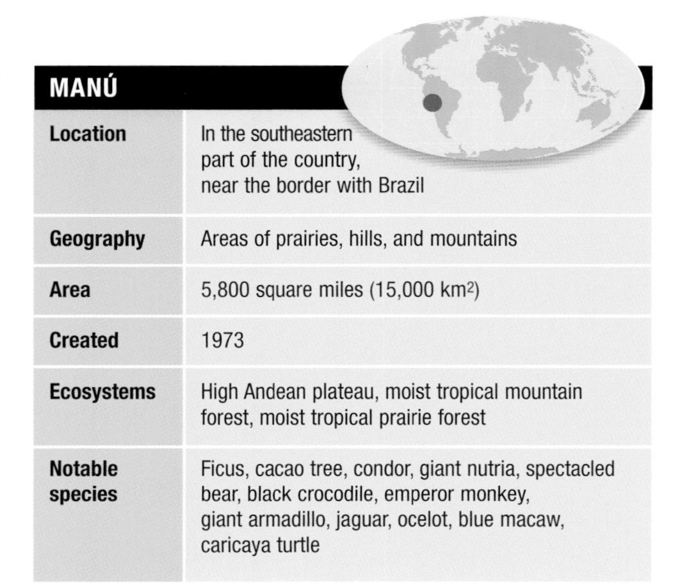

MANÚ	
Location	In the southeastern part of the country, near the border with Brazil
Geography	Areas of prairies, hills, and mountains
Area	5,800 square miles (15,000 km^2)
Created	1973
Ecosystems	High Andean plateau, moist tropical mountain forest, moist tropical prairie forest
Notable species	Ficus, cacao tree, condor, giant nutria, spectacled bear, black crocodile, emperor monkey, giant armadillo, jaguar, ocelot, blue macaw, caricaya turtle

In Manú National Park, some species that are common to other areas of the Earth reach spectacular sizes.

WOOD BUFFALO NATIONAL PARK (CANADA)

This is one of the largest parks in the world and one of the best representatives of the **prairies** of the boreal regions. It contains the largest **inland delta** on Earth, in Athabasca Lake. Here there are herds of bison preserved in perfect condition. There are around 500 species of plants, 227 of birds, and 47 of mammals. The climate is extreme continental, with very cold winters.

WOOD BUFFALO	
Location	The northwestern section of the country, in Alberta and the Northwest Territories
Geography	Great plains, rivers, and lakes
Area	17,300 square miles (44,800 km^2)
Created	1922
Ecosystems	Plains, boreal forest
Notable species	Fir, pine, bison, wolf, lynx, fox, falcon, ptarmigan, great gray owl, snowy owl, Canada goose, diver, crane

Once on the verge of extinction, the bison is now out of danger in the Wood Buffalo National Park in Canada.

BIALOWIEZA NATIONAL PARK (POLAND/BELORUSSIA)

This park is located in a remote area that is difficult to get to, with lakes, swamps, and forests that made it possible for species such as the **European bison** and the **Tarpan horse** to survive. It retains large stretches of unaltered **fluvial forest**. It has a cold continental climate. More than 900 species of higher plants have been counted.

BIALOWIEZA	
Location	Northeast Poland and southeast Belorussia
Geography	Plains, lakes
Area	310 square miles (800 km^2)
Ecosystems	Fluvial forest
Notable species	Spruce, pine, alder, oak, linden, European bison, lynx, nutria, beaver, Tarpan horse, moose, deer, wolf, eagle, grouse, crane

The nutria is one of the most abundant animals in the Bialowieza National Park.

DOÑANA NATIONAL PARK AND RESERVE (SPAIN)

Located in one of the two main **migration** routes for northern European birds headed for Africa, this is also an important **wintering area** for many of these birds. It contains several ecosystems located around the wetlands of the Guadalquivir River. It has a **Mediterranean climate** with a fairly long dry period during the summer. It contains more than 450 species of vertebrates.

DOÑANA	
Location	In the southeastern part of the Iberian Peninsula, at the mouth of the Guadalquivir River
Geography	Floodplains and coast
Area	298 square miles (772 km^2)
Created	1969
Ecosystems	Dunes, wetlands, Mediterranean forest
Notable species	Iberian lynx, mongoose, deer, nutria, fallow deer, goose, teal, flamingo, malvacea, purple gallinule, black vulture, Spanish imperial eagle

Wetlands in the Doñana National Park in Spain, and an Iberian lynx, its most typical (and threatened) animal species.

NIOKOLO-KOBA NATIONAL PARK (SENEGAL)

This park extends along the Gambia River over flat terrain, with lots of **woods**, which allows the existence of great **biodiversity**. This is one of the main natural refuges of Western Africa. Currently more than 1,500 plant species have been identified.

NIOKOLO-KOBA	
Location	The southeastern part of the country, near the border with Gambia
Geography	Practically flat plain
Area	35,250 square miles (91,300 km^2)
Created	1954
Ecosystems	Gallery forest, dry, fluvial forest
Notable species	Bamboo, elephant, buffalo, roan antelope, giraffe, papion, chimpanzee, colobus monkey, lion, leopard, crocodile, hippopotamus, bateleur

Giraffe in Nikolo-Koba National Park.

SERENGETI NATIONAL PARK (TANZANIA)

This park covers a plain that ends in moderately high mountains and is crossed by several year-round rivers. The climate is very hot, and the rains are concentrated in a **rainy season**, followed by a dry season. This park contains the most typical landscape of the African **savannas** and has a great richness of fauna.

SERENGETI	
Location	The northeastern part of the country, near the border with Kenya
Geography	Plains and medium-sized mountains
Area	5,700 square miles (14,700 km^2)
Created	1951
Ecosystems	Savanna
Notable species	Acacia, gnu, gazelle, zebra, hartebeest, desert warthog, giraffe, hyena, lion, leopard, hippopotamus, elephant, wild dog, cheetah, rhinoceros

VIRUNGA NATIONAL PARK (RWANDA)

This is a region of very diverse landscapes from **savannas** and **swamps** to **high mountain areas**, with climates that likewise are very varied and have great differences in rainfall. The great inequality in precipitation favors the existence of several **types of vegetation**. There are many rivers, and this is the wintering ground of many birds from the Northern Hemisphere.

VIRUNGA	
Location	In the northeastern part of the country, on the border with Uganda
Geography	Plains, rivers, hills, 15,500-foot (5,000 m) mountains
Area	30,500 square miles (79,000 km^2)
Created	1925
Ecosystems	Fluvial, savanna, dry forest, mountain rainforest
Notable species	Bamboo, waterbuck, okapi, East African bongo, gorilla, buffalo, chimpanzee, elephant, lion, hippopotamus, francolin, pelican, bontebok

Hippopotamuses in a waterhole in Serengeti National Park (left). Elephant on the giant volcano Ngorongoro, a special preserve in Serengeti National Park (center).

Mountain Gorillas in Virunga National Park.

SUNDARBARNS NATIONAL PARK (INDIA/BANGLADESH)

This park occupies a large area of the mouth of the **Ganges** River, with the largest **mangrove** forest in the world. The greatest altitude does not exceed 33 feet (10 m) above sea level, so the **tides** are continually changing the structure of the emerged areas. The climate is very hot, with plenty of rain in the **monsoon** season.

SUNDARBARNS	
Location	The mouth of the Ganges River, between India and Bangladesh
Geography	Floodplain
Area	3,900 square miles (10,000 km^2)
Created	1984
Ecosystems	Mangrove swamp, estuary
Notable species	Sundari trees, Bengal tiger, fishing cat, axis deer, Bengal macaque, Ganges River dolphin, saltwater crocodile, ibis

Mangrove swamp near the mouth of the Ganges River in Sundabarns National Park.

TE WAHIPOUNAMU PARKS (NEW ZEALAND)

This is a united series of several national parks and preserves that extend 54 miles (90 km) inland from the coast. They offer a great variety of landscapes, with **fjords**, **mountains**, and **volcanoes**. The climate is **oceanic**, with high humidity and precipitation up to 390 inches (10,000 mm) per year. Many of the animals that live here have yet to be studied.

TE WAHIPOUNAMU	
Location	Southeast of the South island
Geography	Very rugged coast, mountains
Area	10,100 square miles (26,000 km^2)
Created	1952
Ecosystems	Coastal, mountain, oceanic forest
Notable species	New Zealand seal, crested booby, brown kiwi, spotted kiwi, kea, kaka

Landscape in Te Wahipounamu National Park.

UJUNG KULON NATIONAL PARK (INDONESIA)

This park includes the famous volcano **Krakatoa**. The landscape is varied, both land and coast. The various types of **forest** shelter a rich fauna, including threatened species such as the Java rhinoceros. The climate is very rainy and tropical.

UJUNG KULON	
Location	Far western island of Java
Geography	Rugged coast with many islands; interior mountainous
Area	470 square miles (1,200 km^2)
Ecosystems	Coral reefs, dunes, tropical forest
Notable species	Podocarpus, palms, Java rhinoceros, dhole, fishing cat, binturong, leopard, gibbons, long-tailed macaque

One of the islands formed by recent eruptions of the volcano Krakatoa, in the east of Java and in the Ujung Kulon National Park.

THE ECOLOGY MOVEMENT

The increasing deterioration of environmental quality in the second half of the twentieth century awakened people's conscience and caused them to demand cleaner conditions for life. This gave rise to what is known as the **ecology movements**, the purpose of which was to reclaim the right of humankind to enjoy unspoiled nature and the right to life of the remaining inhabitants of the planet. The ecology movements became the engine that drove many **social reforms** pertaining to humans' behavior toward nature.

GREENPEACE

This is a nongovernmental agency that has become one of the main ecology groups on the planet. It is financed exclusively by dues from its members (several million throughout the world), so it can remain independent of companies and governments. It uses its funds to carry out spectacular **campaigns** in defense of the **environment** and protests against assaults on nature.

A ship belonging to Greenpeace, an independent association that seeks to maintain the ecological conscience in a world where the main interests are economic ones. To increase its efficiency, it uses national organizations in many countries.

The cruel killing of seals in the Arctic awakened many consciences and spurred the ecology movement.

Since the seals do not recognize the threat of humans, they do not flee from their executioners, who kill them by hitting them on the head with clubs.

All products that come from whales can now be manufactured from other products, so there is no sense continuing to kill these rare animals.

THE WHALING MORATORIUM

In 1994 the International Whaling Commission approved this indefinite moratorium, which prohibits the hunting of these cetaceans.

DEFENDING SEALS AND WHALES

The killing of **seals** in the Arctic regions by cruel methods to get their fur appeared in the communications media and triggered major protests throughout the world. The same thing happened with **whaling**, giving rise to international meetings intended to avoid the extinction of these great cetaceans. Thanks to the actions of ecology groups, which awakened the public conscience, these killings are being stopped, in spite of the commercial interests of the fur and whaling companies.

SEAL HARVESTS IN THE NORTH ATLANTIC

Year	Quota	Official harvest	Estimated harvest
1994	186,000	52,916	264,376
1995	186,000	4,794	258,964
1996	250,000	242,262	508,082
1997	275,000	264,204	499,465
1998	275,000	282,070	532,516
1999	275,000	244,552	498,315
2000	275,000	91,602	337,219
2001	275,000	226,493	484,109

This includes the estimates of Bycatch, Arctic, and Greenland (calculating on the basis of takes published by Stevenson et al., 2000, Walsh 1999, and NMFS 2000)

ANTARCTICA

This virgin continent has been the focus of many defense campaigns. This involves not only saving a virgin territory and avoiding its deterioration but also scientists have presented many proofs that show the great importance of this ecosystem for the whole planet. In the Antarctic waters, especially south of parallel 40, a sanctuary has been created for the reproduction of whales.

The island of Paulet in Antarctica.

The first international conservationist protocol was signed in 1900, in this case to save the African fauna.

POLITICS AND ECOLOGISTS

In many countries, especially in Germany, the ecology movements (also referred to as "**greens**") have received enough votes to participate in governing the nation. In nearly all industrialized countries, mainly in Europe, the participation of ecologists in politics has contributed to raising people's consciousness about the serious environmental problems that affect our world.

The first ecology movements arose at the end of the nineteenth century, but they brought together only small numbers of people.

ACRONYMS

The WWF, or World Wildlife Fund, is one organization with the goal of collecting funds for the conservation of wildlife in all parts of the planet. The IUCN, or International Union for the Conservation of Nature, is an international organization that coordinates conservation activities, mostly governmental, across the entire world.

ALPHABETICAL SUBJECT INDEX